OUTCATS

An "Outcat" is an outcast
and a far-out cat combined.
—the pianist Paul Knopf,
1959

outcats

JAZZ COMPOSERS,
INSTRUMENTALISTS,
AND SINGERS

FRANCIS DAVIS

New York
OXFORD UNIVERSITY PRESS
1990

Oxford University Press

Oxford   New York   Toronto
Delhi   Bombay   Calcutta   Madras   Karachi
Petaling Jaya   Singapore   Hong Kong   Tokyo
Nairobi   Dar es Salaam   Cape Town
Melbourne   Auckland

and associated companies in
Berlin   Ibadan

Copyright © 1990 by Francis Davis

Published by Oxford University Press, Inc.,
200 Madison Avenue, New York, New York 10016

Oxford is a registered trademark of Oxford University Press

LIBRARY OF CONGRESS CATALOGING-IN-PUBLICATION DATA
Davis, Francis.
Outcats : jazz composers, instrumentalists,
and singers / Francis Davis.
p.   cm.
ISBN 0–19–505587–X
1. Jazz musicians—United States—Biography.
I. Title.
ML385.D29   1990
781.65'092'273—dc20
[B]   89–23031   CIP   MN

1 2 3 4 5 6 7 8 9

Printed in the United States of America
on acid-free paper

Again,
for Dorothy Davis
and Terry Gross

For Borah,
who unwittingly
gave me my title

And in memory of
Mark Moses

CONTENTS

EIGHT SINGERS AND A COMIC

COMBOS, MOVEMENTS, ISSUES, AND ISOLATED EVENTS

INTRODUCTION: JUST A MOOD

Which of the jazz performers discussed here are outcats? Obviously, those not in step with any movement or school (like Ran Blake, Borah Bergman, Errol Parker, and the late Herbie Nichols). But also those living in political or culture exile (like Abdullah Ibrahim or Steve Lacy), those operating outside of New York City (Edward Wilkerson, Jr., and the Ganelin Trio), those who remain suspect because of their gender (Jane Ira Bloom and Michele Rosewoman), those with divided loyalty to jazz and pop (Harry Connick, Jr., Susannah McCorkle, and John Zorn), those with drug or alcohol problems (Charlie Parker, Lester Young, Billie Holiday, Frank Morgan, and Sheila Jordan), and those whose genius makes them *sui generis* (Duke Ellington, Gil Evans, Cecil Taylor, Miles Davis, and Sun Ra). But by virtue of the marginal status of jazz in contemporary American culture, all jazz performers, including the most famous, influential, and housebroken, are outcats.

So, too, are those of us who listen to them. I seldom go to parties, but when I do, the people to whom I'm introduced have no idea what they're supposed to say to me when I tell them that I make my living writing about jazz. Talking to me as if I were a fan rather than a writer, they sometimes ask me where's a good place in town to hear jazz. I tell them that the question they *should* be asking is which jazz performers are worth searching out, and the conversation usually ends there. This is probably just as well, because they wouldn't like the performers I recommended, anyway. The alienation that one is likely to feel as a result of one's advocacy of jazz is a leitmotif in this collection, and, no doubt, one explanation for its moody, introspective tone. (Another reason for it is the realization that so many proven musicians have been reduced to playing their horns for small change on street corners and to living in shelters or on the streets—they have become "outcats" for real.) But insofar as alienation is my "theme," it emerged after the fact, which is as it should be. I didn't begin with a thesis, and I don't think that any critic worth reading ever really does. Regardless of his overview, a critic should deal with his subjects on a case-by-case basis, and that is what I've tried to do.

Outcats collects thirty-seven of the profiles and critical essays (a good number combine both modes) that I published in various newspapers and periodicals between 1986 and the beginning of 1989 (or slightly earlier, where noted). It should be read as one critic's assessment of which performers were most worth thinking about—and therefore most worth writing about—during that period. Needless to say, some of the music that struck me as especially provocative during this period was of an earlier vintage. My subjects are arbitrarily grouped: I am aware, for example, that my "Composers" are also "Instrumentalists," and vice versa. The category a performer appears in reflects nothing more that the approach I took to him or her. Certain major figures, including Sonny Rollins, Ornette Coleman, David Murray, and Anthony Davis, are unrepresented because I feel that I've already said all I have to say about them for the time being in *In the Moment*, my first collection. I suppose I could justify the inclusion of the late pop singer Bobby Darin and the political satirist Mort Sahl by pointing to their jazz influence, but a better justification is that they fascinate me for the same reason that many of the strays I'm attracted to in jazz do: they don't fit in. I spend many of these pages bickering with my press colleagues because I feel that jazz journalism is in crisis, with so many of the veteran writers who might be expected to lend the perspective of age having forsaken journalism for administrative positions (or writing about only the kinds of jazz with which they are already familiar), and too many of the younger ones committed only to what's fashionable at any given moment.

I'd like to express my gratitude to Sheldon Meyer and Gail Cooper of Oxford University Press, and to the editors who initially prepared these pieces for publication: William Whitworth, Corby Kummer, Barbara Wallraff, Eric Haas, and Avril Cornel of the *Atlantic*; Linda Hasert, Peter Landry, and Avery Rome, of the *Philadelphia Inquirer*; Milo Miles and John Ferguson of the *Boston Phoenix*; Doug Simmons and Gary Giddins of the *Village Voice*; Georgia Christgau and Ken Richardson of *High Fidelity*; and Don Shewey of *7 Days*. These pieces benefited from their touch. Even so, no piece appears exactly as first published (a perfectionist is somebody who knows that nothing is perfect), and many have been substantially expanded, now that column inches are no longer a concern. Special thanks to Pauline Kael.

Philadelphia
April, 1989

composers

SURVIVING ELLINGTON

There's no business like show business. An *opéra bouffe* that Duke Ellington conceived in leisure—beginning work on it as a vehicle for Lena Horne in the late 1950s, and returning to it periodically, between more urgent deadlines, until his death in 1974—was being executed in haste. Six weeks before *Queenie Pie* was scheduled to open in Philadelphia as part of the American Music Theater's third season in 1986, not a single part had been cast.

"The key to the whole situation is finding our Queenie Pie. Once we know who she is, we'll know who Lil Daddy, her lover, and Café O'Lay, her younger rival, should be," explained Mercer Ellington, a white-haired man in his late sixties, sitting at a table littered with bios and eight-by-ten glossies in a sub-basement of New York's Lincoln Center. The auditions were over, and now Mercer and the others—the musical director Maurice Peress, the choreographer and director Garth Fagan, the co-director Robert Kalfin, the librettist George C. Wolfe, the lyricist George David Weiss, and Marjorie Samoff and Eric Salzman of the American Music Theater Festival—had assembled to judge the "callbacks." Each member of this braintrust was looking for something different, and some compromise would have to be reached between the actors who couldn't sing, the singers who couldn't dance very well, the dancers who couldn't sing or act, and the "gypsies" who did a little bit of everything, although not very well.

Something that no one else had noticed (or, at least, mentioned) was bothering Mercer Ellington. All of the talent parading before the tribunal was black, and that would never do. "Every show my father was ever involved with was racially integrated, and this one will be, too, if I'm going to have anything to do with it," Mercer said quietly, but in a tone of voice designed to indicate that the point was nonnegotiable, looking directly at Kalfin, who had wondered aloud whether white characters belonged in a show set in 1920s Harlem and on an imaginary island in the West Indies. "Ellington wouldn't have wanted anything to do with a log-cabin revue—that's what he would have called this, the way it seems to be going—and I don't want anything to do with it either, if that's what you have in mind."

"Was *Sophisticated Ladies* integrated, Mercer?" asked Peress, peering at Ellington over his bifocals.

"Of course it was! You know that, Maurice," Mercer replied.

"Yes, and look how awful that turned out," Peress quipped. The tension dissolved in laughter, and Elissa Meyers, the show's casting agent, promised to scare up some white faces before nightfall.

"Just make sure they're qualified," Mercer warned her. "I've seen the reverse situation too often in the old days, when agents would send black singers and dancers who were all wrong for the parts, in order to ensure that no blacks would be hired."

Mercer Ellington was listed in the festival's program as *Queenie Pie*'s production coordinator, a job that no one, least of all Ellington himself, could define. The Duke Ellington Orchestra would be the pit band, but Mercer Ellington would not be holding the baton, as he had since assuming leadership of the orchestra upon his father's death. Instead, Peress would be the conductor, which seemed only proper as it was he who worked closely with Duke Ellington on the orchestrations and the vocal score.

Still, everyone agreed that Mercer Ellington was indispensable to the production, even if his role was merely to advise and consent. "He's standing in for the great presence," Peress explained. "He's here to make sure that what happened with *Sophisticated Ladies* doesn't happen again. That show was an entertainment which didn't really give a full account of Duke Ellington's genius. The ideal situation would be to have Duke himself involved in every aspect of the production, but, of course, that isn't possible. So Mercer is here because he's kept the flame burning."

"The first time I heard Ellington's music on the radio, it didn't make that big an impression on me, to tell you the truth. What did impress me was that we even *owned* a radio. You have to remember that this was back in the days when radio was considered an electronic marvel, when our home and the streetlamps outside were still being lit by gas," Mercer Ellington remembered over coffee one morning in his apartment across the street from Lincoln Center. He was due to hear more *Queenie Pie* hopefuls later that day. Three likenesses of his father (a silkscreen, a watercolor by Mercer's daughter Gaye, and the K. Abe photograph that served as model for the Ellington postage stamp) hung on one of the living room walls. Otherwise, there was surprisingly little Ellington memorabilia (his manuscripts are in a sealed vault in the Irving Trust, along with the Medal of Freedom bestowed on him by President Nixon), and unless one happened to notice the small electric piano next to the ironing board in the hallway corner, there was no apparent evidence that the man of the house was himself a musician and composer.

For anyone enamored of Duke Ellington's music, an audience with his son is a thrill, because no one else living knows that music so intimately. Even so, no one would want to trade places with Mercer Ellington, because more than anyone else, he also knew the man behind the music, and theirs was an unusually adversarial father–son relationship. Still, Mercer seems to have survived relatively unscarred, with his sense of humor intact—his reminiscences are punctuated by laughter, sometimes rueful, but never really bitter. Dressed haphazardly, as he was that morning, in knit shirt, track pants, and sneakers, he conveyed none of his father's seignorial air. Although he has a smoker's cough, he didn't once reach for his cigarettes during the two hours or so I spent with him. He generally refers to his father as "Ellington"—a sign, perhaps, that the debonair bandleader kept his son at the same distance that he kept the rest of the world (as well as a tacit admission by Mercer that the name "Ellington" inevitably evokes Duke, not Mercer).

"He called me 'the brat,' and I addressed him as 'Fadu,' for Father Duke, until I was about twenty. After that, I called him Pop, although we mostly talked to each other without calling each other anything. Ironically, although I always knew he was popular, even revered, it only fazed me how truly important Ellington was in the music business during one of his low periods, around 1950, when big bands were folding right and left, including mine. I had a mortgage to pay off, and three children to send through school. Borrowing money from Pop was out, because he wasn't doing so well himself. I would ask other bandleaders to give me a job, they'd say 'Why should you collect a paycheck that someone else really needs? You're Duke Ellington's son.' Everybody assumed he was rich, and he might have been if not for the problem of back taxes and keeping his men on salary, come rain or shine, as the saying goes, forty-eight weeks a year. If he hadn't been so determined to keep his band, his song royalties would have made him a millionaire several times over."

Duke Ellington never went out of his way to make the road easier for his son. If anything, he put hurdles across the path. "Duke Ellington would make certain he remained on top regardless of whom he knocked down, including me," Mercer wrote in his 1978 memoir, *Duke Ellington in Person* (written with the jazz critic Stanley Dance), an uneasy mix of *Daddy Dearest* and *Father Knows Best*. If only one Ellington could have his name in lights, Duke wasn't going to let it be Mercer, who remembered in his book that when he formed his own band in the 1940s, Duke insisted that he join him on the Musicraft label, then saw to it that only the band's inferior material was released. According to Mercer, Duke feared that *two* Ellington bands would confuse the public and devalue the name.

Mercer Ellington was born in 1919 in Washington, D.C., the first child of Edward Kennedy and Edna Thompson Ellington, who were both under twenty-one. Mercer doubts that his parents would have married except to legitimize his birth. "There was too much pulling them apart. My mother's folks were from a higher station of black society than my father's. They were all schoolteachers and principals, and they considered all musicians, including Duke Ellington, low-life. Ironically, though, Ellington's parents thought my mother wasn't good enough for *him*. You know how parents are, and his were even more so."

Two years later, Duke and Edna had another son, who died in infancy. So Mercer grew up as an only child, but without the unlimited devotion usually lavished on only children. He was left behind with his paternal grandparents in Washington when his parents set out for New York to launch Duke's career. His parents sent for Mercer when he was nine, but separated soon after, and although they never divorced, they were never reconciled, either. "I came home from school one day, and there was a strange woman [the dancer Mildred Dixon, the first of Duke's many paramours] living with my father and taking care of me and Ruth [Duke's younger sister, only four years older than Mercer, to whom she was more a sibling than an aunt]. My mother, it turned out, had moved back into the building up the street, where we had lived before. They had separated without telling us. Nobody in my family liked to be the bearer of bad news."

Until he entered college, Mercer spent six months a year with each parent, an unusual custody arrangement for the period. "But it wasn't a legal arrangement, you see. It was just something I decided on my own," says Mercer, who remained close to his mother until her death in 1966. What kept drawing him back to his father was their mutual love of music, although, ironically, it was his mother—a good friend of Fats Waller's—who gave him his first piano lessons, showing him the chords to Duke's "Solitude." "When I was staying with her as a kid, I would have to shine shoes, or whatever, to bring some extra money into the house. She was a lonely woman who liked to entertain just to have company, and although my father supported her to some degree, her friends would drink up every penny that she had. In a way, it's good that Ellington wasn't more generous with her, or she might have drunk herself into an early grave, with those friends of hers around."

Mercer also enjoyed a warm relationship with Beatrice "Evie" Ellis, a former Cotton Club showgirl who became Duke Ellington's mistress in the late thirties, and whose death followed his by two years (she is buried beside him in New York's Woodlawn Cemetery). "Poor Evie," Mercer says. "She was typical of many attractive black women at the Cotton Club; very intelligent but not very well educated. The goal for these women was to snare a handsome black bandleader, because if they fooled around with the white patrons, the best they could hope to become would be mistresses. Evie latched on to Ellington, who refused to divorce my mother and had no intention of remarrying after she died. So Evie wound up as a mistress anyway. She felt very exploited, and worried that she wouldn't be taken care of if he died, because he didn't believe in wills. She acquired a reputation as a grump, which enabled Ellington to use her as a shield. If he invited someone over and then changed his mind, he'd send her to the door to chase them away. The next time he ran into them, he'd act like Mister Sunshine and ask them why they hadn't shown up, giving the impression that she kept people away from him because she wanted him all to herself.

"You try to give your own kids what you never had when you were growing up, and, in my case, that was a feeling of closeness to my father. That's why I've always tried to take my kids on the road with me, whenever possible. I remember when *Sophisticated Ladies* opened on Broadway in 1980, I was sipping champagne at the celebration party with Paul, my youngest, wrapped up in my arms. He was only a year old then—you see, I wanted a grandchild, and since none of my kids were making any progress in that direction, I decided to take matters into my own hands." (In addition to Paul and teenage stepson Ralph, Mercer has three grown children from his first marriage: daughters Mercedes, a dancer, and Gaye, a painter; and son Edward Kennedy Ellington III, until recently the guitarist in the Duke Ellington Orchestra. "So, altogether, I have five children, although some people think I have six when they see me with Lene," Mercer jokes, referring to his second wife, a Danish airline hostess who pronounces her name "Lena" and is a good twenty-five years Mercer's junior.) In *Duke Ellington in Person*, Mercer recalls that sight of himself in uniform during World War II triggered an uncharacteristic show of fatherly affection from Duke. He can now guess what went through his father's mind. "I never knew fear until my first child was born. The most horrifying thing I can imagine is to outlive my children."

. . . The photographers cocked their shutters and snapped the picture, Duke happily looking paternal, right index finger pointing in the air as he admonished Mercer.

"Now grow up," he told Mercer, "and be a great composer," Duke laughed.

"As you say, Father Duke," Mercer smiled back.

Barry Ulanov, *Duke Ellington* (1946)

Although Duke Ellington studied art in high school and was offered a scholarship to Pratt Institute, he was musically self-taught, with an autodidact's scorn for formal education. Yet he saw to it that Mercer studied music at Juilliard, Columbia, and New York University. "I think it was his way of keeping up with advancements in music theory through me," Mercer speculates. "I remember one day he handed me two huge volumes and said, 'Read these and tell me about them.' It was the Schillinger System, and it took me three years to digest it. When I started to explain it to him, he cut me off. 'Oh, yeah, I was doing that back in 1928.' And, truthfully, he had." Predictably, Mercer's greater education deepened the rift between them. "He valued the intuitive approach, and I'm belatedly coming around to his way of thinking. Because unless you count the hit version of 'Bill Bailey' I did for Della Reese, which was just an arrangement of a song in the public domain, the only things I've written that've had any lasting impact were all written for the Ellington Orchestra *before* I went to music school," Mercer laughs, referring to "Things Ain't What They Used to Be" and "Blue Serge," which are often mistaken for his father's work. "I take it as a compliment when people assume that something of mine was written by him. That's given me tremendous confidence."

Mercer, a competent musician, was a disappointment to his father, a genius who, in terms of professional encouragement treated Billy Strayhorn—a composer of uncommon ability, only four years older than Mercer—more like a son. Still, Duke never banished Mercer from the fold, and Mercer showed little inclination to leave. In 1964, after being on the fringes of his father's organization for two decades, Mercer joined the Duke Ellington Orchestra as its road manager and a member of its trumpet section. "As road manager, I was a combination psychologist, mathematician, and private detective. In order to get everyone on the bus or plane in the mornings, I had to keep track of who everybody had shacked up with the night before. With Paul [Gonsalves], who had a drinking problem, although it never interfered with his playing, I'd have to look for him on the floor of his hotel room if he wasn't in his bed. Or, once or twice, in the alley outside the hotel. Paul was a special challenge to my father, which is why he loved him so much. The guys prided themselves on being an unruly bunch, and since each man had been responsible for his own hotel bills, the band had acquired an undesirable reputation for skipping out the back windows without settling its accounts. I changed all that by taking care of all the travel expenses myself and deducting the costs from each man's salary every week. I felt my father had a dignified image that had to be maintained. What made the job really difficult was that I was riding herd on the men who had helped to raise me: Cootie Williams, Johnny Hodges, Harry Carney. When someone knows you as a kid, it doesn't matter how much you age. You'll always be the brat to them. It's like spending your whole life in your hometown. Did they resent me telling them it was time to go to work? You bet they did. They wouldn't do what Ellington told them to do, so why should they listen to me? That's why he had this opening vamp that he would play that eventually wound up being copyrighted as 'Kinda Dukish.' The idea was for the audience to think that the band was waiting for Ellington to make up his mind what tune he wanted to play first. But he was actually waiting for the stragglers to take their chairs."

Would Duke Ellington ever second-guess his son? "Not really, because he didn't want to be bothered with the sort of details I was handling for him, if that meant taking away time from composing. But once I almost got into trouble for giving Russell Procope a five-dollar-a-night raise without consulting him. Russell was the lowest-paid man in the band, even though he had been with us for years. I liked him very much, so I snuck the extra money into his check without saying anything. After about three weeks, Russell said to my father, 'I've been getting too much money. What am I supposed to do? Give it back?' To my surprise, Ellington said, 'Well, Russell was honest about it, let's leave it that way.' Afterwards, I said to Russell, 'You know, if this were any other band, you could have cost me my job. You weren't giving me a chance to save my own skin by giving me the opportunity to correct what could have been an error on my part. From now on, come to me if you have any problems with your paycheck.' Of course, Russell became my enemy after that."

Did Duke ever compliment Mercer on a job well done? "No, and as a matter of fact, if anyone else would start to praise me in his presence, he would immediately shush them or change the subject, afraid that I would ask him for a raise. But he showed his appreciation in other ways. He would call me on the phone in my hotel room, ostensibly to argue about one thing or another, but really just to talk. I think he beckoned me mostly to keep him company, though he was too stubborn to admit that. We knew each other too well by that point for him to expect that he was getting a yes-man. From time to time, he would ask me if I loved him, and I would tell him that I did."

Thankless as the job sounds, it at least enabled Mercer to spend his father's last decade by his side, an opportunity given few sons. "That was important because I had never had the chance to know him that well at home. He was always on the road. Mine had always been the gruesome job in the family—I had to bury the relatives while Ellington worked. And when he was diagnosed as having cancer, there was the gradual shock of realizing that I would be burying him. Even before Ellington died, there were times when we would be on a bill with singers, and it was understood that we were to back them up. And he would call on me to lead the band so that he could save his energy for his own set. He also insisted that the band fulfill its contractual obligations even when he was in the hospital. He had promised the president of IBM that the band would fly down to Bermuda for an IBM convention with or without him. So a few days after his funeral, I was on a plane for Bermuda. Churches all over the country were inviting the band to play his Concerts of Sacred Music, and I gradually found myself holding a baton instead of a trumpet."

Mercer Ellington has been a semipro football player, a disc jockey, and a traveling salesman as well as a musician. He gives the impression that he would be just as happy behind the scenes, except for one thing: "Ever since I walked out on stage to take a peek at the audience when I accompanied Ellington on a New England tour when I was around seven years old, I've wanted to be on stage from time to time. I used to tell my mother's folks that I was interested in a career in aeronautics, but that was just to appease them. I always wanted to be a musician. That's why I learned both trumpet and saxophone as a kid. I wanted to be ready for whatever vacancy opened up in Pop's band. I never dreamed of leading his band. I just wanted to be in it with him."

"He passed away about a week before we went to Bermuda for IBM, and I discovered when we got there that we were in danger of losing our way. Paul Gonsalves was dead. Harry Carney was ailing. Cootie [Williams] was planning to retire. There were new men coming into the band, and the men who knew his music by heart were on the way out. We had only about twelve arrangements written down on paper for the new men to learn, and maybe a dozen others we could fake. What I had to do was restore the library, by hiring arrangers to transcribe older pieces from records, including "Caravan" and "Perdido," both of which were licks that guys in the band had come up with that just grew and grew, like Topsy, but had never been written down. Through the process, I began to catch up with him. I'm still catching up with him. But it's too big a job to accomplish in my lifetime, which is why it's important that his original arrangements be written down."

Under Mercer's direction, the Ellington Orchestra has revived "Ko-Ko," "Birmingham Breakdown," "East St. Louis Tootle-oo," "Hot and Bothered," "Daybreak Express," "Ring Dem Bells," "Echoes of Harlem"—innovative works from the twenties, thirties, and forties that Duke himself neglected in later decades. These reinterpretations leave much to be desired. The ensemble blend is usually too brassy, and in the absence of Hodges, Carney, Gonsalves, Lawrence Brown, and Cootie Williams, the solos are nondescript. Moreover, record companies, when bringing the Duke Ellington Orchestra into the studio, have inevitably requested new versions of Duke Ellington's Greatest Hits: "Caravan," "Satin Doll," "Sophisticated Lady," and "Take the 'A' Train." This has been a source of frustration for Mercer. So was *Sophisticated Ladies*, although with this show he succeeded in giving his father the Broadway smash that eluded him during his lifetime.

"When I was in the army, we produced what we called 'blueprint' shows, which meant that you went from base to base with just the score for a show—no cast. So the songs were tailored for the average serviceman to sing. What we would do is decide which songs should be in the show, then lay the sheet music on a table and keep shuffling it around until the song titles suggested a story. I preached the idea to Pop of doing the same thing with his songs when I got out of the service. I said, 'Your track record of unsuccessful shows is what they're holding against you. You should make your success as a songwriter work for you by putting all of your best-known numbers in one show.' But it was no sale. He said that would be cheating. He wanted each of his shows to contain all new songs specifically written for it." *Sophisticated Ladies* was to have been a "blueprint" show. "Let the songs tell a story. I got the idea again from seeing my daughter Mercedes in *No, No, Nanette*. It was such a pleasure to sit there hearing songs that were already familiar. But that was the last time [the show's producers] listened to me."

Although Mercer was listed as music director and conducted the Duke Ellington Orchestra for most of *Sophisticated Ladies'* Broadway run and many of the subsequent revivals, he had no creative control. "What I objected to about the show was that there was a bit too much Broadway in it. The music wasn't the star of the show, the way it should have been. The dancing and the staging took precedence. The best singers in the cast, like Priscilla Baskerville, were given the least to do, and they often had to sing at tempos all wrong for them, and all wrong for my father's music, because of the dancers, who argued that the tempos should be suited to *them*, since they were the ones risking injury. My argument was, make the dance fit the music, instead of the other way around. That's what Alvin Ailey had always done with Ellington's music. We wound up compromising on the issue, but nobody was happy. Every night, there would be either a singer or dancer waiting to chew me out.

"What I liked about *Sophisticated Ladies*, though, was that it presented Ellington to a contemporary audience. It meant that an additional generation would be familiar with his work. But, you know, every time he had a big hit song, he felt that that permitted him the luxury to do something major which might not go over so big. I feel that one of my obligations is to expose every side of Ellington, and that's why *Queenie Pie* is so important."

Queenie Pie didn't become one of Duke Ellington's priorities until 1970, when New York's WNET-TV commissioned its completion. Two years later, Ellington recruited Maurice Peress—former maestro of the Kansas City Philharmonic and the Corpus Christi Symphony, and once Leonard Bernstein's understudy with the New York Philharmonic—to assist him in orchestrating the vocal parts.

"The television people did not renew their option after Duke's death," Peress explains, "presumably because they felt the piece wasn't valuable without his presence on camera as the narrator, which is really what they were paying for." Patti LaBelle and Robert Guillaume (TV's "Benson") sang a few numbers from the score on *Love You Madly*, a 1983 PBS special. But it looked as though *Queenie Pie* would never be performed in its entirety—until Peress brought Mercer Ellington together with Marjorie Samoff and Eric Salzman of the American Music Theater Festival.

Queenie Pie had "American Music Theater Festival" written all over it in large letters—as a lavish amusement by a serious composer who doubled as a pop songwriter (like Gershwin's *Strike Up the Band*, which the festival had revived in 1984), and as a jazz composer's attempt to hurdle the racial and class prejudices segregating jazz from the other performing arts (like Anthony Davis's opera *X*, which the festival had debuted that same year). "We fell in love with the idea," Samoff says. Still, *Queenie Pie* was hardly ready to be staged, although Ellington had left among his personal effects what appeared to be a finished score, a libretto, and a plot synopsis neatly typed by Betty McGettigan. ("Another of his girlfriends, with a literary gift," Mercer says. "He believed in putting them to work.") The show wasn't long enough at sixty minutes (the length of the proposed TV presentation). So Peress interpolated "Creole Love Call," as well as four new songs culled from Ellington's notebooks and given lyrics and dramatic rationale by George David Weiss and George C. Wolfe, respectively.

The title character was loosely based on Madame C. J. Walker, a key figure in the Harlem Renaissance of the 1920s, a beauty-products entrepreneur believed to have been the first woman in the United States to earn a million dollars. But Ellington's scenario was pure fantasy. As the show opens, Queenie Pie has reigned as Harlem's leading cosmetician and most beautiful woman for longer than anyone can remember. Suddenly facing stiff competition from a conniving young arrival from New Orleans, she sails off in her yacht and washes ashore on an imaginary island where she learns of a magic elixir that will bestow the eternal youth she needs to retain her throne.

Those involved with the American Music Theater production saw *Queenie Pie* as a metaphor for Ellington's own soul-searching during his final years. Peress says, "My sense of it is that like a lot of famous people who have been everywhere and done everything, Ellington still felt somehow unfulfilled." Samoff says, "He was looking at the success of the Beatles and coming to terms with his own mortality, although I don't want to make the opera sound too heavy, because so much of it is tongue-in-cheek. But it deals with a mature and successful person coming to terms with a younger generation, asking the question, What is timeless and what is ephemeral?" Mercer Ellington says, "What [Queenie Pie] is really searching for [is] love, although she's not even aware of this need."

But this sounds like post–Phil Donahue pop psychology. Ellington made his peace with the Beatles with a delicious arrangement of Lennon and Mc-Cartney's "All My Lovin'" on *Ellington '66*, and a backbeat gave him nothing to fear—witness "Acht O'Clock Rock" from *The Afro-Eurasian Eclipse* and the irresistible "Blue Pepper" from *The Far East Suite*, both from the mid-sixties (to say nothing of "Happy Go Lucky Local," from 1946). And would Ellington have depicted a feminine alter ego, given what his son describes as "a basic contempt for women. He considered them a necessary evil. The only women he ever had anything good to say about were Ruth and his mother, both of whom he put on a pedestal." More likely, Ellington's conscious theme was vanity, which he thought of as a feminine trait—although it was really a family trait, according to Mercer, who grew up calling his paternal grandparents Aunt Daisy and Uncle Ed so they wouldn't feel old. "Ellington was the same way. You can imagine how he felt when I made him a grandfather. When reporters would assume that I was his brother, or that my children were his, he would never correct them."

Would Duke have been happy with the American Music Theater Festival's production of *Queenie Pie?* For that matter, was Mercer happy with the way it was shaping up? "I argue with Bob [Kalfin] and Garth [Fagan] that since there are very few people in the theater who can sing, dance, and act equally well, we should go with the singers. But I have to honor their decisions. I believe what Ellington would have done is to wait until all the parts were cast, then overhauled everything to suit the abilities of the performers, just like he did with his orchestra. Whenever something turns out differently than you first expected it to, you know you're being faithful to the spirit of Duke Ellington," says

Mercer, who, by his own admission, has spent the last twelve years trying to think exactly like his father, after spending his first fifty-five years doggedly thinking just the opposite.

And so, on with the show.

Is *Queenie Pie* major Ellington? On the basis of a gala preview I attended, the answer would have been an unequivocal no. The jerry-built sets inspired by Romare Bearden's background scrim of Harlem in the 1920s (itself something any hack magazine illustrator might have come up with) looked like they would topple, and so did the dancers, in the oversized period costumes they wore in the first act. The choreography was a dire mix of jazz, ballet, and *Solid Gold*. Worst of all, Ellington's melodies had been only sketchily orchestrated. The show was inert, and everyone knew it. When an AMTF staffer with a good memory and a theatrical flair passed me on her way back to her seat after the intermission, she looked me in the eye and paraphrased a line from *Tea and Sympathy*: "When you write about this—and you will—please be kind."

But following a return visit a week later, I was no longer so sure that *Queenie Pie* was hopeless. A lot of hard work had gone into the production, and it showed; although the book—as dated and patronizing in its assumptions about "primitive" cultures and what women really want as Ellington's earlier *A Drum Is a Woman* (another TV project)—was dead weight no amount of effort was going to salvage. But perhaps it was a mistake to take the story too seriously. Just as some of the trills that Ellington wrote for his title character affectionately mocked Verdi and Puccini, it's possible that the ludicrous scenario was a send-up of classic opera's farfetched and superfluous story lines. (Queenie's real-life model made her fortune selling pomade to black Americans who wanted straight hair, but Ellington—or Wolfe—reversed the pattern by having Queenie show island folk how to let their hair grow natural. She introduces them to the concept of soul. Talk about American know-how!)

In any case, the book hardly mattered. The story was merely scaffolding for the music, which was majestic, although one would never have guessed that early in the show's run. From sumptuous balladry of the kind that casual listeners most associate with Ellington ("Truly a Queen"), to the minor-key mysterioso he bequeathed to Thelonious Monk and other contemporary jazz composers ("My Father's Island"), to the modal, percussive exotica that dominated the later suites ("Stix"), every conceivable genre and subgenre of Ellingtonia was represented in the score. Knowing that Peress, a staunch Ellingtonphile, had left the production over "creative" disagreements, I feared the worst; but the Ellington Orchestra, under Roy Glover's direction, succeeded in bringing to the surface melodies that had been hopelessly submerged during the previews. In most cases this involved stripping away excess orchestration. But "My Father's Island," which had earlier pivoted on a minimalistic organ vamp, had been reharmonized for brass and reeds, and now sounded royal. It was fitting that the members of the orchestra took their places alongside the singers and dancers during the curtain calls. They were among the stars of the show, and they deserved a hand.

"Truly a Queen" and the three other previously unpublished Ellington songs that Peress and Wolfe had interpolated in the score fit comfortably, even if Weiss's lyrics were frequently banal (not surprising, given that he wrote the words for "What a Wonderful World" and "Can't Help Falling in Love," dreck hits for Louis Armstrong and Elvis Presley, respectively). Throughout *Queenie Pie* there were references to earlier Ellington, including the opening fanfare from *Harlem*, impeccably played by the trumpeter Barry Lee Hall to bring down the final curtain. "Jam with Sam" became "Harlem Scat." "Rhumbop" was reprised from *A Drum Is a Woman*. Only the inclusion of the wordless "Creole Love Call" seemed a misjudgment, though not because of its earlier vintage. Patty Holley gave it a boffo delivery that stirred up pleasant memories of Adelaide Hall on the original 1927 recording. The only problem was that Holley was playing Queenie's rival, Café O'Lay, and Teresa Burrell as Queenie had no comparable first-act showstopper to win the audience over to her side. After seeing the preview, I wrote Burrell off as a graduate of the Phyllis Hyman School of Soulfulness and Charm. But once she gained assurance, she turned out to be a diverting singer with a voice full of weird dissonances ideally suited to Ellington. Larry Marshall and Ken Prymus were amiable enough in the male leads, and the energetic Lillias White and the Billy Eckstine–like baritone and former Ellington band singer Milt Grayson almost stole the show in minor parts (she took the lead vocal on "Stix," the second act's big dance number, and he sang "Oh, Gee," a *Sophisticated Lady* offspring). The entire cast deserved applause for scrupulously avoiding the pop-eyed gesticulation of other black shows set in Harlem of yesteryear, including the original production of *Sophisticated Ladies*.

After its Philadelphia premiere, *Queenie Pie* played a month at the Kennedy Center in Washington, D.C. A year later, George C. Wolfe, still flush with the success of *The Colored Museum*, told the *New York Times* that he was doing research on the Harlem Renaissance in preparation for bringing *Queenie Pie* to Broadway; so far, nothing has come of this. The staging and dancing still need work, to judge from the Philadelphia performance, but the music is a fine starting point. *Queenie Pie* shouldn't be allowed to die. At the very least, someone should sponsor the Ellington Orchestra in a concert presentation of the music. And perhaps it's not too late for a cast album.

But I'm begging the original question: Is *Queenie Pie* major Ellington? If Ellington were still alive, it wouldn't matter. The best numbers would go into his working repertoire, where they would undergo further revision, and the others would be forgotten. But the problem with posthumous discoveries is that they have to be masterpieces or ephemerae, with no allowance for anything in between. *Queenie Pie* is neither. Ellington came closest to integrating music and spectacle in his three concerts of sacred music, his sound track for Otto Preminger's 1959 film *Anatomy of a Murder*, and (presumably) his Cotton Club revues. Compared to these, *Queenie Pie* is Ellington Lite. But *Queenie Pie* affirmed that even minor Ellington can be pretty wonderful, and suggested that the moment was overdue for revivals of *Jump for Joy* (1941), *Beggar's Holiday* (1947), and *My People* (1963), as well as first performances of *Poussé Café* (1962) and the undated *The Man with Four Sides*.

(*September* 1986)

ELLINGTON'S DECADE

Jazz scripture insists that composition is, at best, a springboard to improvisation; at worst, an obstruction to it. But jazz scripture also insists on progress, and in the past decade, as improvisers have been retracing their steps, composers have been the ones breaking new ground—writing formally ambitious works that, while not eschewing improvisation altogether, relegate it to second place and demand more rigorous self-editing by improvising soloists. Many of these composers are rejecting jazz's traditional isolationism to collaborate with poets, choreographers, and classical instrumentalists. What was unique about X, Anthony Davis's opera about the life of Malcolm X, which was presented by the New York City Opera in 1986, was that it actually got funded—there could be other X's waiting to happen, by Dave Burrell, Ornette Coleman, Abdullah Ibrahim, and Henry Threadgill, all of whom have described similarly grand initiatives. Other contemporary composers thinking big—if only in terms of thinking orchestrally, even when they are able to hire just six or seven players— include Muhal Richard Abrams, Anthony Braxton, John Carter, Joseph Jarman, Leroy Jenkins, Steve Lacy, Roscoe Mitchell, Butch Morris, James Newton, Errol Parker, George Russell, Leo Smith, Cecil Taylor, Edward

Wilkerson, and the members of the World Saxophone Quartet (Hamiet Bluiett, Julius Hemphill, Oliver Lake, and David Murray) and the String Trio of New York (Billy Bang, James Emery, and John Lindberg). But if one accepts the premise that composition has been the key element in jazz in the eighties, the decade's central figure is not any of these, but Duke Ellington, who died in 1974.

Ellington's influence has never been greater. Typed as a "jazz" composer only by circumstance of race, he spent his career chafing against the restrictions of jazz, much as his spiritual descendants are chafing now. His scope was enormous. In addition to ballads even shapelier and riffs even more propulsive than those expected of a swing-era big-band leader, his legacy of more than fifteen hundred published compositions includes tone poems, ballet suites, concertos for star sidemen, sacred music, topical revues, film scores, picture-postcard-like impressions of faraway places, and extended works unparalled in jazz until very recently and classifiable only as modern American music. (His only shortcoming was as a popular songwriter, and, in part, this was because he never found his Ira Gershwin or Lorenz Hart.) A butler's son who dared to imagine himself the lord of the manor, an experimentalist who courted and won mass acceptance, Ellington was one of America's greatest composers (regardless of idiom), and perhaps the most quintessentially American in the ease with which he navigated the distance between the dance floor and the concert hall.

In 1984, the jazz critic Gary Giddins estimated that fifty hours of new Ellington recordings had been released in the ten years following his death. More have been released since then, including long-forgotten studio sessions and concert tapes previously circulated only among collectors. Invaluable as much as this material has proved to be, it is ironic that it has generally been easier to come by than Ellington's most celebrated work—the epochal sides he recorded for RCA Victor in the early forties, for years available only on French import labels or by mail order from the Smithsonian Institution.

The 1986 release of *Duke Ellington: The Blanton-Webster Band* (RCA Bluebird 5659-1-RB, also available on cassette and compact disc) indicates that RCA is finally beginning to realize what treasures lie in its vaults. The four-record set collects the Ellington Orchestra's entire commercially released output from March, 1940, to July, 1942—arguably Ellington's most fertile period, although most of his larger-scale works, beginning with *Black, Brown, and Beige*, were still to come. By 1940, most of Ellington's sidemen had been with him a decade or longer, and he had been so important in shaping their sensibilities that he could virtually predict the content of their improvisations.

(Indeed, in some cases he wrote their "improvisations'" out for them before-hand.) This enabled him to take daredevil risks as a composer and arranger, and gave even his "through-composed" works an improvisatory ring. The two newcomers alluded to in the collection's title perhaps stimulated him even more. Jimmy Blanton, the first jazz bassist to phrase with a horn player's melodic fluidity, gave the Orchestra's syncopations unprecedented bite and opened Ellington up as a pianist. The arrival of Ben Webster, the Orchestra's first tenor saxophone star, gave Ellington another ace soloist to call on, as well as another color for his palette. (He acquired still another, more exotic color in 1941, when Ray Nance, who doubled on violin, replaced Cootie Williams in the trumpet section.)

The detail of Ellington's writing and the individuality of his soloists never fail to astonish, no matter how familiar one is with the tracks on *The Blanton-Webster Band*. "Concerto for Cootie," with its beautifully elongated theme and Cootie Williams' piquant variations on both muted and open horn, is a masterpiece that lost something when Ellington rounded off the main theme from ten bars to eight to accommodate Bob Russell's lyrics, and the piece became "Do Nothin' Till You Hear from Me." "Cootie" is here in its pristine form, and I envy anyone hearing it for the first time—the same goes for the surging dialogues between Blanton and the full ensemble on "Jack the Bear." "Ko-Ko" offers intimations of pedal-point modality and *Kind of Blue*; and the delight of "Cotton Tail" (based on "I Got Rhythm," in anticipation of bebop) lies in its layered sax-section riffs and Webster's sinuous choruses (which foreshadowed both the "tough" tenor style and Coltrane's "sheets of sound"). The charms of "I've Got It Bad (And That Ain't Good)" include Ellington's celeste, Paul Francis Webster's plain-spoken lyrics (the finest ever to grace an Ellington song), the lovely countermelody between choruses, and the blissful conjunction of Johnny Hodges's alto and Ivie Anderson's voice (as harmonious a union as that achieved more informally a few years earlier by Lester Young and Billie Holiday). "Main Stem" and "Harlem Air Shaft" are among the tracks that highlight Ellington's penchant for experimentation even within the conventions of the blues and the thirty-two-bar popular song format. The numerous cover versions of pop ditties of the period ("Chloe" and "You, You, Darling" are my personal favorites) demonstrate Ellington's powers of transfor-mation, even though Herb Jeffries's ungainly warbling proves that even Ellington was human (with the exception of Anderson, he never employed a first-rate singer on a regular basis). This was also the period in which Billy Strayhorn, Ellington's protégé, blossomed into an influential composer and orchestrator under Ellington's watchful eye. Strayhorn's "Raincheck" and "Johnny Come Lately" anticipate bebop phraseology, and his "Take the 'A'

Train," which ultimately became the Ellington Orchestra's signature theme, cleverly underlines the band's playful swank. But the most evocative Strayhorn piece here is "Chelsea Bridge," with its lordly Webster solo—as a successful jazz appropriation of Ravel and Debussy, this remains unsurpassed even by Ellington, a master impressionist in his own right.

It's too bad that the producers of *The Blanton-Webster Band* failed to include Ellington's 1941 duets with Blanton, or the small-group dates from the same period led by Hodges, Bigard, and the trumpeter Rex Stewart, all of which featured Ellington on piano. If it is true, as Strayhorn is said to have put it, that Ellington's real instrument was the orchestra, it's equally true that the piano became an orchestra at his urging. (*Money Jungle* [Blue Note BT-85129], a bristling encounter with the modernists Charles Mingus and Max Roach, recorded in 1962 and reissued with additional material in 1986, provides a good, long look at Ellington the dissonant stride pianist.) Although the vintage performances on *The Blanton-Webster Band* have been digitally remastered, the sound isn't as vivid as on the French reissues, nor is the pitch always accurate. If RCA intends to do justice to its Ellington catalog, its job is far from over—the Ellington Orchestra recorded masterpieces for the label before 1940 and after 1942. *The Blanton-Webster Band* is a godsend for those on a tight budget: others are advised to search the specialty shops for French RCA's increasingly difficult-to-find *The Works of Duke*, twenty-four volumes available separately or in five boxed sets. Still, music that is timeless and universal in its appeal belongs in chain stores as well as in specialty shops, which is why the reappearance of this material on a well-distributed domestic label is so welcome. *

In the years since Ellington's death, iconoclastic performers associated with the jazz avant-garde have recorded albums of his compositions. They have brought their own agendas to his music, which has proved more malleable than anyone might have imagined (though there is some justice in the complaints of those who insist that to play Ellington means playing him his way or not at all). The most striking of these revisionist homages are the flutist James Newton's *African Flower* (Blue Note Bt-85109), the pianist Ran Blake's *Duke Dreams* (Soul Note

---

* *The Blanton-Webster Band* ends with titles recorded on July 28, 1942, after which a musician's strike against the record companies kept the Ellington Orchestra out of the Victor studios until December 1, 1944. *Black, Brown, and Beige* (RCA Bluebird 6641-1-RB), another four-record set, released in 1988, resumes the chronology (and the sound is much better than on *The Blanton-Webster Band*). Included along with *The Perfume Suite* and excerpts from *Black, Brown, and Beige* are "Midriff" and "Esquire Swank," two 1946 performances inexplicably omitted from the otherwise complete *Works of Duke*.

SN-1027, distributed by PolyGram Special Imports), and the World Saxophone Quartet's *World Saxophone Quartet Plays Duke Ellington* (Nonesuch 79137-1), the most recent of these entries, as well as the most wholly satisfying in terms of fealty to Ellington's tempos and the ineluctable *rightness* of its deviations from text.*

That so many performers who are generally adamant about playing their own original material would choose to interpret Ellington is eloquent testimony to his inexhaustible influence, as are the Ellingtonian flourishes (sometimes filtered through his disciple Charles Mingus) that pervade the writing of John Carter, Abdullah Ibrahim, David Murray, and Henry Threadgill. But when I call Ellington the key figure of the eighties, it's not just because modernists continue to play his tunes or to aspire to his orchestral majesty. Musicians from all stylistic camps have long done that much, and Mercer Ellington keeps his father's most familiar music in circulation with albums like *Digital Duke* (GRP GR-1038), released in 1987. (If anything, the familiarity of the material on *Digital Duke* is a disadvantage; with so much obscure Ellington and Strayhorn still awaiting discovery, new versions of "Satin Doll" and "Take the 'A' Train" are luxuries we can live without.) Nor is it just because plungered, speechlike brass styles like the ones that Bubber Miley, Cootie Williams, Rex Stewart, and Joe Nanton patented during their years with Ellington are again all the rage, thanks to the trombonist Craig Harris and the trumpeters Lester Bowie and Olu Dara. Nor is it because Anthony Davis and many other black composers are consciously, and in some instances programmatically, giving musical expression to the goals and frustrations of black society, just as Ellington did with such mural-like works as *Harlem*, *The Deep South Suite*, and *Black, Brown, and Beige*. This is Ellington's decade because visionary jazz composers are taking up his unfinished task of reconciling composition and improvisation. Jazz is thought of as extemporaneous and fleeting—that's another part of its romance—but these composers are aiming for a perpetuity like that which Ellington achieved. They are mounting larger works, as he did, without worrying whether the results strike everyone as jazz.

Of course, not everyone who listens to jazz is as sanguine about this development as I am. Some fear that jazz is recklessly heading for the same dead end that classical music arrived at earlier in this century with atonality and serialism. And it's true that the composers I nominate as Ellington's heirs lack his common touch, his willingness to play the role of entertainer; and they will probably never find themselves in a position to develop this commendable trait.

---

* Chico Freeman's *Tales of Ellington* (Black Hawk BKH-537-1) should be added to this select list.

Jazz has experienced growing pains since Ellington's time, and the innocent idea of entertainment had become enslaved by the cynical science of demographics (the question no longer is whether something is entertaining, but how many, and precisely *whom*, it entertains). In its maturity—some would say its dotage—jazz has become an "art music"; because a reconciliation with pop seems out of the question, a rapprochement with classical music is probably the key to its survival. Ellington gives these contemporary composers much to strive for, but his mass appeal is out of their reach. This is unfortunate, because a larger audience should be part of what a composer hopes for when he starts to think big.

(August 1987)

POSTSCRIPT

"Ellington's Decade," which was written on assignment for *The Atlantic* in January, 1987, finally ran in the August issue (as "Large-Scale Jazz"). Several readers found it curious that my piece didn't address James Lincoln Collier's Ellington biography, which—entirely by coincidence—appeared in stores that same month. They were correct in assuming that I disliked the book.

Collier makes two serious charges against Ellington. The first—that he "stole" from his sidemen—is easily dismissed. As Gene Santoro pointed out in an article on Miles Davis in *down beat*: "Ellington set up what was essentially a closed feedback loop between himself and his band members. He kept his players on salary and used the studio and bandstand as compositional sketch pads. If during the course of a workout or jam a Johnny Hodges or a Tricky Sam Nanton or a Juan Tizol hooked a melodic idea that grabbed the Maestro, he'd seize it (sometimes with credit, sometimes not) and weave his rich orchestral tapestry around it. In the process, it would become a total composition, something living in a fully ramified way, quite apart from the scrawl or flash that had given it birth. . . . [This] helps [to] explain why so few of Duke's sidemen ever went on to become imposing bandleaders in their own right, even aside from the very real socioeconomic considerations involved." By Collier's logic, *Rhapsody in Blue* was written, not by Gershwin, but by Ross Gorman, whom Gershwin is supposed to have heard playing that famous clarinet glissando during a Paul Whiteman band rehearsal.

The second charge—that because Ellington's music depended so heavily on improvisation and so little of it was put down on paper, "we are entitled to question not just whether [he] was America's greatest composer but whether he was a composer at all"—is more difficult to debate, based as it is on an apparently un- shakeable value judgment. In applying the same highbrow standards that he elsewhere mocks Ellington for "naïvely" aspiring to, Collier is competing with his subject, and rigging the rules so that Ellington can't win. "Duke was finding ways to slide by that sounded to lay audiences a good deal better than they would to professionals," Collier muses at one point; and at another, "I am certain that Duke would not usually have been able to analyze his own scores in the way that a more formally trained arranger would." So what? Dismissing the *Sacred Concerts*, Collier notes that although "Duke's fans and the loyal jazz press always found good things to say about [them] . . . I have not been able to find support for this position among *professional* music critics" (italics mine). It takes a few readings to realize that Collier is referring to the hacks who review classical music for daily newspapers, all jazz critics presumably being well-meaning amateurs who are sometimes lucky enough to be paid for their opinions.

Only a fool would rank the *Sacred Concerts* among Ellington's greatest works, but the point is that classical reviewers generally took a patronizing tone toward Ellington when they paid him any mind at all. The goal of a biographer should be to put his subject in perspective; Collier just wants to put Ellington in his place. Although Ellington's hypochondria, superstitions, active libido, and often-callous behavior toward those closest to him provide the ammunition for what Joyce Carol Oates calls "pathography," Collier lacks the rancid eloquence and prurient curiosity of an Albert Goldman or an Arianna Stassinopolous Huffington—his *Duke Ellington* is neither as vile nor as readable as their Lennon and Picasso bios. But even when praising Ellington, Collier's tone is grudging, and some of his comments about Duke and his intimates are so mean-spirited and condescending as to be construed as unconsciously racist. "The first of Duke's rules was to break them," Collier writes, and slipping into his characteristic amateur psychoanalytical mode, adds, "He was driven to this conclusion by temperament, not by any careful process of thought," as though it made a difference. Collier's Ellington is a man-child of impulse whose disdain for formal training was a sign of "laziness" rather than the justified arrogance of the autodidact. As portrayed by Collier, Ellington's father was a silly old Kingfish whose determination to become a "gentleman" was an expression of "pretentiousness," and whose "feeling for the elegant"—food and wine and china and such—was "superficial" emulation of the wealthy white family he worked for. Why can't Collier give this self-educated black man the benefit of the doubt? The grown men in Ellington's orchestra are more than once referred to as "boys."

In subjecting Billy Strayhorn to special abuse, Collier plays off of Strayhorn's alleged homosexuality: "Ellington always evinced a tendency—weakness, if you will—toward lushness, prettiness, at the expense of the masculine leanness and strength of his best work, the most 'jazzlike' pieces. Strayhorn encouraged this tendency. . . . To be sure, Strayhorn wrote some very nice swingers . . . but in the main he was exploring a tropical rain forest thick with patches of purple orchids and heavy bunches of breadfruit. Duke's work increasingly moved in this direction." In other words, Strayhorn *feminized* Ellington's music. Collier stops short of suggesting that Strayhorn, not music, was Ellington's mistress.

This book poisons the waters for the rest of us. I doubt I am the only critic who has found himself overcompensating on Ellington's behalf. (In a review of Strayhorn homages by Art Farmer and Marian McPartland in the *Philadelphia Inquirer*, for example, I crossed out a line about Strayhorn's being Ellington's superior as an impressionist for fear that readers would conclude that I agreed with Collier that Ellington wasn't all he was cracked up to be.) Collier has already been criticized strongly by, among others, Santoro in *The Nation*, Dan Morgenstern in the *Times Book Review*, Martin Williams in the *Washington Post*, Michael Ullman in the *Boston Phoenix*, and Gary Giddins in the *Times Literary Supplement*. But the estimable Michiko Kakutani's favorable review in the *New York Times* demonstrates how tragically little America's best and brightest know about jazz, and how easily Collier's distortions could become the Official Version. Kakutani, whose specialty is minimalist fiction, can be forgiven for repeating Collier's half-truths. But what excuse is there for E. J. Hobsbawn's rave in *The New York Review of Books?* Forget that Hobsbawn used to write jazz criticism under the name of Francis Newton. As a socialist, he should be able to recognize cultural conservatism.

(*August 1987/February, 1989*)

HOTTENTOT POTENTATE

> What distinguishes American heroes of this kind [those of Cooper, Melville, and James] . . . is that there is nothing in the real world, or in the systems which dominate it, that can possibly satisfy their aspirations. Their imagination of the self . . . has no economic or social or sexual objectification; they tend to substitute themselves for the world. Initially and finally at odds with "system," perhaps their best definition is Henry James Sr.'s description of the artist as hero . . . "the man of whatsoever function, who in fulfilling it obeys his own inspiration and taste, uncontrolled either by his physical necessities or his social obligations." The artist-hero may be, as he often is in American literature, an athlete, a detective, or a cowboy. . . .

Or an autotheistic bandleader in his early- to mid-seventies who lords it over his sidemen (many of whom are themselves getting up in years) in a Philadelphia rowhouse. The quote from Richard Poirier's *A World Elsewhere: The Place of Style in American Literature* describes not only Sun Ra's indifference to temporal realities (he says he wasn't born of a woman, and, according to what his tenor saxophonist John Gilmore told Bob Rusch in *Cadence*, he doesn't plan on dying, either), but also his independence from any jazz movement or school.

Ra is obviously a major figure in jazz, even if it is difficult to pinpoint why. He's often assumed to have been the spiritual father of the Association for the Advancement of Creative Musicians, given the circumstantial evidence of his Chicago roots, his theatricality, and the multi-instrumentalism he demands of his musicians. But Ra and true believers kept to themselves in Chicago in the fifties, and to my knowledge, no AACM charter member has ever cited him as a progenitor. His relationship to free jazz was provisional, if not antagonistic—the Arkestra's group-gropes were meant to represent the chaos that would ensue on earth without Ra's divine intervention. Despite his prescient use of electronics, his music bears little resemblance to fusion or Ornette Coleman's harmolodics. He's frequently hailed as the last of the great big-band leaders, but despite the Duke Ellington and Fletcher Henderson tunes in his repertoire, his loyalty to that tradition is theoretical. Even when able to afford a full horn section, he hasn't exploited its potential—a big-band *sound* is less important to him than the blueprint for black solidarity he understands big bands to embody.

Decades from now, Ra's career might be divided into three tidy phases on the basis of his albums (most likely excluding the hundreds he's released on his own now-you-see-them, now-you-don't, Saturn label). It will seem as though the Sun Ra Arkestra started off as an anomalous, hard-bop big band with regrettable exotic trimmings in the no-nonsense fifties, mutated into an avant-garde stunt troupe in the freaked-out sixties, and finally hit its stride as a ragtag repertory orchestra in the time-warped eighties. But records are deceptive, especially in Ra's case. At Sweet Basil or the Bottom Line, you never know which Arkestra you'll get: with luck, you'll get all three. In substituting his panoramic overview for whatever happens to be in fashion at any given moment, Sun Ra has become doggedly *sui generis*, and part of the fun in hearing him is the recognition that there has never been anyone so intransigent or bizarre. In terms of the marketplace, he is a lowly cult figure, but his avid following is surprisingly broad-based. In addition to jazz buffs, he appeals to new-music types looking for their own crackpot-genius *à la* the art world's Howard Finster; to the metallists, iron men, and Lester Bangs wanna-be's who write for the *Village Voice*; and to his own equivalent of the Deadheads—pot smokers old enough to remember when the Arkestra travelled with a light show, who still feel free to light up at Ra's concerts. (Given the camp extravaganza of his stage presentation and rumors about his sexual preference, it's only surprising he doesn't have an identifiably gay following.)

The only jazz musician assured of a gig every Halloween, Ra sports a planet-dotted tunic, sequined knit hat, tie-dyed goatee, and a patriarchal smile on the cover of *Reflections in Blue* (Soul Note BSR 0101). As the first album to justify claims for Ra as a latter-day Fletcher Henderson, *Reflections in Blue* is noteworthy for more than Ra's iconic cover pose. The program includes no Henderson, just Irving Berlin and Jerome Kern as Henderson might have played them, and originals with reed voicings as luxurious as Henderson's or Don Redman's, but with enough personal twists to let you know this is Sun Ra.

The swing arrangements that began to dominate Ra's sets around ten years ago confirmed suspicions that he was as much showman as shaman, that his love of pomp had more to do with race memory of hot jazz and hokum, the Cotton Club and the TOBA circuit, than with ancient Egypt and outer space, his cosmological mumbo-jumbo notwithstanding. But the benefits of those all-night drills that Ra is rumored to put his men through aren't always evident, and once the shock of hearing the Arkestra tear into "Queer Notions" or "Yeah, Man" wears off, the band's chancy intonation becomes a problem. What's intended as sincere homage frequently sours into parody. Yet it was difficult to condemn Ra for this shortcoming, because his willingness to present virtuosos and rank amateurs side by side was what validated him as an avant-gardist in the sixties and early seventies, when a desire for cultural identity, rather than a fascination with chords and scales, motivated many black musicians to pick up horns relatively late in life. (One agenda of most postwar avant-garde movements, including free jazz, has been to show that master craftsmen have an awful lot to learn from motivated beginners.)

Still, it comes as something of a relief that the pros have regained the upper hand on *Reflections in Blue*. The best of the new recruits are the guitarist Carl LeBlanc, the trombonist Tyrone Hill, and the trumpeter Randall Murray, all of whom solo energetically and add spark to the pep sections below the solos. But the album's MVP is homecoming alumnus Pat Patrick, once the Arkestra's baritone saxophonist and now its lead altoist (assuming it's he and not Marshall Allen lofting those gorgeous lines). Patrick's swiftly articulated solos on "Yesterdays," Ra's "State Street Chicago," and his own "Nothin' from Nothin'" (the album's catchiest theme) are delightful for laying anachronistic bop licks over Ra's charleston, shuffle, and two-beat rhythms. Allen turns in leaping choruses on the title track (more a modified boogie-woogie than a blues, and one of Ra's earliest numbers). John Gilmore, the Arkestra's star soloist from the beginning, although not featured as prominently as one might wish, performs his usual chordal prestidigitation on several tracks. Ra, playing synthesizer as well as piano, evokes Ellington, Basie, Garner, Monk, and Meade Lux Lewis, often simultaneously. He sings "I Dream Too Much," taking his cue from the oneiric title rather than the lovelorn lyrics, and giving a whole new meaning to the word *atonal*, although the result is very endearing.

Clearly recorded and well distributed (by PolyGram), *Reflections in Blue* is as definitive as Sun Ra albums get. Manny Farber, the film critic, once made a distinction between two kinds of art: "white elephant," which lavishes "overripe technique" on hammy, self-aggrandizing pseudo-masterworks; and "termite," the subversive Farber's own preference, characterized by "buglike immersion in a small area without point or aim, and . . . concentration on nailing down one moment without glamorizing it, but forgetting this accomplishment as soon as it has been passed; the feeling that all is expendable, that it can be chopped up and flung down in a different arrangement without ruin." Sun Ra, Hottentot Potentate of a World Elsewhere, is his own white elephant, but most of his carelessly produced concert albums are those of a termite artist. A *Night in East Berlin*, originally released as a Saturn cassette but now available as Leo LR-149 (from New Music Distribution Service) fades in on what sounds like the middle of a number. Without visual cues, it's impossible to make sense of what's going on on the remainder of side one, especially after Ra brings things to a standstill with his Phantom of the Opera synthesizer.

But despite dismal sound quality, side two is a reminder of how much fun Ra can be. After the processional "Interstellar Loways" (a cross between "Tenderly" and Ra's own "Fate in a Pleasant Mood," with an elegant piano intro and a fluent alto solo), Ra paces the Arkestra and the June Tyson Dancers through a medley of his greatest hits, including "Space Is the Place," "We Travel the Spaceways," "Rocket Number Nine," "Next Stop Jupiter," and "Shadow World" (the ultimate showpiece for Gilmore's sustained upper-register squeals—a "Flying Home" for moderns). What's surprising is how much of the rollick of Ra's swing-era adaptations has rubbed off on his own vintage material. Either that, or this stuff is now old enough to inspire its own nostalgia.

No survey of Ra's recent output would be complete without mention of *John Cage Meets Sun Ra* (Meltdown MPA-1). Recorded live at Sideshows by the Seashore in Coney Island in 1986, the album is damned near unlistenable, but fascinating as hell for what it reveals about the differences between the jazz and classical avant-gardes (the title is something of a misnomer; Cage and Ra perform together only briefly). Cage is so important a thinker that it no longer matters what his music sounds like, or whether he bothers to make music at all: the idea is what really counts. As putative entertainers, jazz musicians are denied such luxury, which might be as good an explanation as any for why Ra's sloppy, melodramatic synthesizer solos possess a vitality lacking in Cage's rigorous, empty words (from *Roaratorio: An Irish Circus on Finnegan's Wake?* I can't tell, and Howard Mandel's otherwise informative liner notes don't say). Jazz demands that even its stargazers keep both feet firmly planted when showtime rolls around. For Sun Ra, the show never ends. The audience just goes home.

(February, 1988)

THE INDIVIDUALISM
OF GIL EVANS

Although Gil Evans is widely regarded as the leading jazz orchestrator after Duke Ellington, his career has been mottled with the ironies, paradoxes, and inconsistencies more typical of minor cult figures. Born in Toronto in 1912, Evans is a year older than Woody Herman, and only two years younger than

Artie Shaw—and with his stooped shoulders, thatched white eyebrows, and pouchy mouth, he's beginning to show his age. But his harmonic values, which owe more to bebop and the French impressionistic composers than to swing, distance him from his own generation. Evans—who was pushing forty when he initially made an impression on his peers, nearing fifty when he attained a modicum of popular recognition, close to sixty when he formed a permanent ensemble, and over seventy when he wrote his first film score—was a slow starter who might never have gotten started at all without two shoves from Miles Davis.

The first was in 1948. At that time, Davis and Evans must have seemed like odd running mates. Davis, still in his early twenties, had just completed an apprenticeship with Charlie Parker, the foremost exponent of bebop, then still a radical new form of jazz delighting in the subversion of pop-song conventions. Evans had been an obscure staff arranger for the Claude Thornhill Orchestra, a dance band famous for dreamy ballads like "Snowfall" and "A Sunday Kind of Love," but admired by the beboppers for its translucent voicings, which were made possible by an unusual instrumentation that included French horn and tuba in addition to the conventional brass and reeds. Shortly after leaving Parker's group, Davis formed his own nine-piece band. Although the Miles Davis Nonet was smaller than Thornhill's group and derived its phraseology and rhythmic push from bebop, its instrumentation and palette resembled Thornhill's—it, too, boasted a French horn, tuba, and vibratoless reeds. Significantly, one of the Davis Nonet's arrangers was Evans, whose orchestrations of Charlie Parker compositions for the Thornhill Orchestra had a few years earlier proposed just such a synthesis of bebop abandon and Thornhillian reverie.

The 78s that the Miles Davis Nonet recorded for Capitol in 1949 and 1950 (later collected on an LP titled *The Birth of the Cool*, now out of print) were recognized as masterpieces only in retrospect. In part this was because Davis quickly abandoned this measured approach for an earthier, more casual improvisatory style dubbed "hard bop" long after the fact. This suggests that the *sound* of the Miles Davis Nonet (its most unusual characteristic) was more a reflection of Evans's sensibility than of Davis's. Evans arranged only one of the twelve numbers recorded by the Nonet, and collaborated with Davis on another. Yet the arrangements by Davis, John Lewis, Gerry Mulligan, and John Carisi were rife with what would soon become recognized as Evans trademarks—the shadow-play of horns scored across sections, oppositions of low brass and high reeds, harmonic plasticity bordering on modality. Like Duke Ellington, who is often erroneously credited with writing Billy Strayhorn's "Take the 'A' Train" and Mercer Ellington's "Things Ain't What

They Used to Be," Evans had mastered the genius's trick of creation by proxy. After those early sessions with Davis, Evans retreated into anonymity, until Davis summoned him forth again almost a decade later. From 1957 to 1959, Evans arranged and conducted three albums for Davis that rival Ellington's showcases for Cootie Williams and Rex Stewart as fully realized concertos for trumpet and orchestra. On the strength of *Miles Ahead* (Columbia CJ 40784), *Porgy and Bess* (Columbia CJ 40647), and *Sketches of Spain* (Columbia CJ 40578), Evans began to finish second only to Ellington in the composer/arranger category of jazz magazine year-end popularity polls.

But "arranger" is too limited a designation for Evans, and "composer" might be too grandiose. Evans has never been a prolific composer, and none of his pieces has entered the standard jazz repertoire. Yet it's unlikely that Kurt Weill's "My Ship" (or even Dave Brubeck's "The Duke," a number indigenous to jazz) would have become an enduring jazz favorite without Evans's intervention on *Miles Ahead*. The wistful countermelody with which Evans lightened and modalized George Gershwin's "Summertime" on Davis's *Porgy and Bess* album has since become part of the composition, as far as many jazz musicians are concerned, although one wonders how many of them realize that they are paying respect (but not composer's royalties) to Evans as well as to Gershwin. Much the same is true of the tempo-whipping fanfare that Evans devised to separate Davis's solo from John Coltrane's on an epochal 1956 recording of Thelonious Monk's "'Round Midnight." The anxious, floor-pacing vamp with which Evans introduced John Benson Brooks and Harold Courlander's ballad "Where Flamingos Fly" on a 1957 date for singer Helen Merrill, and again on a 1961 date under his own name recently resurfaced (unacknowledged) as the intro to the guitarist Stanley Jordan's version of John Lennon and Paul McCartney's "Eleanor Rigby." And it would be difficult to imagine "Spain," pianist Chick Corea's best-known composition, without the working model of Evans's arrangement of Joaquin Rodrigo's *Concierto de Aranjuez*, from *Sketches of Spain*. (Evans reintroduced to jazz the rumba, martial rhythms, and Andalusian colorations that Jelly Roll Morton characterized as "the Latin tinge," although in Evans's case, it was more of a Latin twinge—*Guernica*, the Inquisition, and *Death in the Afternoon*.)

In one sense, what Evans does is analogous to what a great improviser like Sonny Rollins does when he bestows his favors on Tin Pan Alley. The difference, of course, is that much of Evans's "improvisation" takes place beforehand, on score paper or in his imagination. Like Rollins, he tends to isolate and expand discreet elements of his source materials—an especially provocative passing chord, a rhythmically insistent vamp, an unexpected melodic elipsis—and discard the rest. As a result, his interpretations have the aura of newly imagined works, ontologically independent of their original inspirations. The best illustration of this is his 1963 reworking of Kurt Weill's "The Barbara Song" (on *The Individualism of Gil Evans*, now available only on compact disc, Verve 833 804 2), on which he gets Barbara's fragrance, but not her scent—her eroticism, but not her disgust at finding herself "per-pen-dic-u-*lar*." By using essentially only the song's bridge, he transforms the composer's anarchic, beer-garden staccato into a spent reverie that is more typical of Evans than of Weill. As an interpretation, it's all wrong, but no less brilliant for that.

Gil Evans's influence on younger jazz orchestrators rests on his association with Davis. It can be argued that his albums with the trumpeter—which were commissions, after all, despite their collaborative nature—misrepresented Evans, suggesting that he was a formalist, an advocate of brevity and compression in a period in which other jazz composers and arrangers were joining improvising soloists in recklessly exploring the possibilities opened up by the advent of the long-playing disc. By 1970, when Evans finally launched his own big band after several unsuccessful earlier attempts, his music had left perfectionism behind in a quest for greater expansion. His later output has suffered from laxity, and it isn't merely a result of his performances having become lengthier, for several of the pieces on *Sketches of Spain* were lengthy by any definition, and the twelve short selections on *Miles Ahead* dovetailed with each other to form an album-length suite. Frequently characterized as an exacting conductor around the time of these albums with Davis, Evans has since become alarmingly laissez-faire. It was often said of Ellington that he could communicate what he wanted from his orchestra with no more than a shrug or a nod: Evans sometimes seems too laconic to make even those gestures. Big bands have traditionally been characterized as either soloists' bands (the Count Basie Orchestra of the late 1930s is the most celebrated example) of arrangers' bands (the Basie band of the late 1950s will suffice). The Gil Evans Orchestra is an anomaly: a big band led by a great arranger, but ruled by its soloists, which, given their prolixity, may be leaving far too much to chance.

Even so, few bandleaders offer as much as Evans does at the top of his game, as he is for much of *Gil Evans and the Monday Night Orchestra Live at Sweet Basil* (Gramavision 18-8610-1), recorded in a New York nightclub in 1984. This two-record set has its share of bloat, in the shape of Alan Shorter's "Parabola" and a Charlie Parker blues medley set to a funk rhythm, each of which consumes an entire LP side without benefit of thematic material varied enough or improvised solos meaty enough to justify such elongation. But the remaining two sides are richly rewarding. It hardly matters that the raw material of Herbie Hancock's "Prince of Darkness" is minimal: Evans's orchestral swirls create the illusion of an ingratiating melody out of a series of harmonic vacillations, and there is a nicely paced solo by Lew Soloff, the latest in a long line of trumpeters whom Evans has used as Miles Davis surrogates. On Charles Mingus's "Orange Was the Color of Her Dress, Then Silk Blue" (or "Orange Was the Color of Her Dress, Then Blue Silk," as it is mistitled here), featuring the tenor saxophonist George Adams, Evans's barrelhouse accelerandos and lusciously vocalized reeds illuminate the erotic turbulence that quickened all of Mingus's ballads (and Adams' solo has a yearning quality that he never displayed with Mingus himself). Evans's treatment of "Goodbye Pork Pie Hat," Mingus's eulogy for Lester Young, is slightly less successful, because the featured soloist (the alto saxophonist Chris Hunter) relies too heavily on double-timed bebop licks—knowing which soloist to call on is apparently no longer one of Evans's strengths. Two numbers are reprised from Evans's 1974 album of Jimi Hendrix compositions. As reimagined by Evans, "Up from the Skies" is a pleasant étude for the entire band, with especially comely brass figures. But the more ambitious "Voodoo Chile" loses its punch about midway through, and Hiram Bullock's Hendrix-like guitar improvisation fails to conjure the spirit of the late rocker, whose charm was in the disparity in scale between his all-stops-out guitar style and his mumbled, affectless vocal delivery. Earlier on, though, there is wonderfully comic braying by Howard Johnson on tuba, along with some atmospheric synthesizer mewling by Pete Levin. Just the raucous tone of the performance is gratifying, in light of the unintentional gentility most jazz musicians bring to rock 'n' roll. Johnson, who plays both brass and reed instruments, is the ensemble's most valuable member—except for the leader, who plays what he once self-effacingly described as "cheerleader" piano.

Like all jazz performers of his longevity and standing, Evans is forced to compete with his past, in the form of reissues. Magnificent though it is in places, *Live at Sweet Basil* suffers by comparison to *Out of the Cool* (MCA/ Impulse 5633), a just-reissued album from 1961. *Out of the Cool*'s most noticeable flourish is "La Nevada," a simple four-bar minor-to-major riff that swells and contracts for almost sixteen minutes behind a succession of soloists, including the trumpeter Johnny Coles, the bass trombonist Tony Studd, the guitarist Ray Crawford, the bassist Ron Carter, and the tenor saxophonist Budd Johnson. "La Nevada" is clearly the prototype for performances like *Live at Sweet Basil*'s "Parabola," but what rescues it from monotony is Elvin Jones's and Charli Persip's varicolored percussion. Over the last two decades, Evans has borrowed from post-Beatles pop, often to obvious advantage: synthesizers, in giving him so much color to work with, have allowed him to become a big-band leader without having to shoulder the expense of ten or eleven horns. But the regimented dance patterns of rock and funk, instead of giving his music the forward motion one might have expected, have brought it closer to the rhythmic standstill it has long courted with its slowly dissolving harmonies.

Still, it might be Evans's recent efforts, not his classic albums with Davis, that are misleading. Evans's all-stops-out adaptations of Mingus's "Boogie Stop Shuffle" and "Better Git It in Your Soul" and his brass arrangements for the pop singers Sade, David Bowie, Ray Davies, and Jerry Dammers, among others, paced one of 1986's most entertaining films—*Absolute Beginners*, British director Julien Temple's hyperkinetic, wildly uneven paean to fifties youth culture and CinemaScope musicals, which was a box-office sensation in England but a flop in the U.S. Avoid the U.S. soundtrack release, a single album excluding Evans's instrumentals, in favor of Virgin VD 2514, a two-record British import. (Evans is also credited with helping to orchestrate Robbie Robertson's score for Martin Scorsese's *The Color of Money*, but his hand is nearly undetectable.) To recommend the *Absolute Beginners* soundtrack over *Live at Sweet Basil* would be perverse. The soundtrack played much better on screen than it does on record, where the absence of improvisational development works against it as jazz. But *Absolute Beginners* at least has greater focus than one has come to expect from Evans lately, and this may be because it was recorded in the studio, like the albums with Davis and unlike the many in-concert LPs that Evans has released since 1978.

If I owned a record company (to indulge in a futile game that jazz critics never tire of playing), I'd bring Evans back into the studio, where greater focus is a prerequisite, and where arrangers are in their element. I'd bring his band into the studio, if that was what he wanted, but I'd also take the opportunity to relieve him of the burden of leading a band. I'd let him pick the soloists for an entire album of Mingus, Ellington, Billy Strayhorn, Kurt Weill (for whom he has long shown a special affinity), or even Stephen Sondheim, whose compositions have been strangely overlooked by jazz musicians, but whose spiraling harmonies seem tailor-made for the Evans treatment. Assuming that a full-scale reunion with Davis is out of the question and that he has no interest in recreating Thornhill, I'd team Evans up with the guitarist James "Blood" Ulmer, who shares his fascination with Hendrix and whose own albums have suffered from unimaginative settings. Needless to say, I'd also bring Evans into the studio for projects of his own choosing, and I wouldn't mind at all if somebody who really does own a record company were to beat me to it. Evans's concert recordings, though uneven, demonstrate that he has maintained his creativity remarkably late in his career. The studio is where that creativity can best be fulfilled.

(February, 1987)

POSTSCRIPT

I guess what I was trying to say, but stopped myself from saying, was, "Won't someone please bring Gil Evans back into the studio before he dies?" Well, someone did, under unusual circumstances that proved to be ideal. *Collaboration* (EmArcy 834-205 2) is essentially a remake of Evans's and the singer Helen Merrill's 1956 album, *Dream of You*. In jazz, remakes are suspect, if not taboo, but this is one instance in which the remake is superior to the original, which itself was far superior to all but a handful of jazz vocal albums of the period.

In 1956, Merrill was an up-and-coming singer with two albums to her credit—one of them a soon-to-be-classic small-group date with Quincy Jones arrangements and Clifford Brown trumpet solos. For her third album, Merrill wanted Gil Evans arrangements, a request that Bob Shad, her producer, initially nixed because, as Merrill explains: "Gil had a reputation for running over budget, for doing most of his arranging on studio time." This presumably explains why Evans had, to that point, never been hired to arrange an entire album, although he had done charts for a variety of singers, including Marci Lutes, Lucy Reed, and even Johnny Mathis. Ultimately, Merrill got her way, and it's good that she did, because without her midwifery, we might have been deprived of *Miles Ahead, Porgy and Bess,* and *Sketches of Spain*. Miles Davis, who had just signed with Columbia and suddenly had a major-label budget at his disposal, decided that the time was right for another project with Evans after hearing Merrill rave about the *Dream of You* sessions on a tour bus.

Evans rewarded Merrill's faith by presenting her in three flattering settings, each employing an unusual instrumentation. "People Will Say We're in Love," featuring what Merrill calls "the hot band" (four brass, two saxes, and rhythm), was an inspired example of orchestrated bebop, complete with "Donna Lee" allusions. Four woodwind-dominated tracks introduced the shaded tonal palette that was to be associated with Evans in the 1960s. But the most stunning tracks, given *Dream of You*'s vintage, were the four ballads with strings, which approached art song without pretention, and included such graceful, unexpected touches as a Bartók paraphrase amid the French impressionist shimmer on "I'm a Fool to Want You." It's unfortunate that Evans was never given another opportunity to write for strings, because on *Dream of You*, he elicited from them a sound more vocal in pitch than rival jazz orchestrators of the period got from actual wind instruments. But despite its virtues (and its tangential Miles Davis connection), *Dream of You* excited little fervor, and remained an obscure collector's item until its 1986 reissue on a four-record box set of Merrill's complete 1950s recordings for Mercury and EmArcy.

We also have Merrill to thank for *Collaboration*. She says that she first approached Evans about doing another album in 1985. Happily consenting, he promised new arrangements, which his failing eyesight prevented him from delivering. Finally, EmArcy's Kiyoshi Koyama (who co-produced *Collaboration* with Merrill) suggested redoing the charts from *Dream of You*. Although initially taken aback at the idea, Merrill eventually decided, "Why not? Classical musicians do it all the time." (So, for that matter, do jazz musicians in concert—but Merrill and Evans had never performed *Dream of You* outside of the studio.) Discounting Evans's soundtracks and his busy work for Davis and Sting, *Collaboration* marked his first studio venture in more than a decade. The sessions, in August, 1987, also marked Evans's final time in the studio as an arranger. (His last date of any kind was *Paris Blues*, an album of duets with the soprano saxophonist Steve Lacy, recorded four months later.)*

Merrill says that Evans told her that he loved his wild, unruly big band, but wanted to hear "color and quiet sounds" again; she resolved to help him raise money to record an album of his own compositions as a follow-up to *Collaboration*. It's too bad that never happened, but *Collaboration* is a triumphant last hurrah. It reprises eleven of twelve numbers from *Dream of You* (studio time ran out before the twelfth—Eubie Blake's "You're Lucky to Me"—could be recorded), and adds "Summertime," with Merrill singing the melody more or less straight, Steve Lacy echoing her one moment and anticipating her the next, and trombone and woodwinds playing a lilting, minimalistic vamp lifted from Evans's arrangement of the same tune for Davis. ("Gil rightly felt that the phrases he created were his compositions," Merrill says, "and he felt free to borrow from himself.") There are slight differences between the two albums, most noticeably the lower keys on *Collaboration*, and the darker, more restless moods the songs take on as a result. (Merrill's voice is darker and less mint-scented than it used to be, but the keys weren't lowered only in deference to her. She told Terry Gross, of the National Public Radio program *Fresh Air*, that the lower tones also suited Evans, who was sick and

---

* *Collaboration* was advertised as Evans's last studio album with a big band. But it turns out that in November of 1987, a young French bandleader named Laurent Cugny invited Evans to Paris to conduct, arrange for, play keyboards in, and record with Big Band Lumière, an ensemble which—to judge from the superb *Rhythm-a-ning* (EmArcy 836-40102)—did a better job of staying within Evans's scores than his own band generally did.

feeling "insulted by age.") Mel Lewis, the drummer on *Collaboration*, is given more leeway than Joe Morello, his counterpart on *Dream of You*, and his brush work is a model of sensitive accompaniment. Digital recording permits the bassist Buster Williams a greater parity with the violins and horns than mere high fidelity permitted the great Oscar Pettiford; and this is no small improvement, given Evans's fascination with the bottom notes of chords (his arrangement of "I'm Just a Lucky So and So" resembles an orchestrated bass solo). Lacy's solo and ad-libbed obbligato on "Anyplace I Hang My Hat Is Home" are new, and both add immeasurably to the performance. The other solos amount to a draw: the trumpeter Shunzo Ono is no match for the lyrical Art Farmer on "By Myself," but the trombonist Jimmy Knepper is a big improvement over the somewhat mechanical Jimmy Cleveland on "People Will Say We're in Love." Evans's later music may not have been as gossamer as his earlier work, but it had a gusto that naturally spilled over into these new interpretations of thirty-year-old charts. You can hear it on "Anyplace I Hang My Hat Is Home," which brings together boogie-woogie rumble and existential dolor so perfectly and so un-self-consciously (with piano and bass vamps and train-whistle flutes) that you think the song was written by Meade Lux Lewis and Jean Paul Sartre, not Harold Arlen and Johnny Mercer.

But none of this is what I have in mind, exactly, when I claim that *Collaboration* is "better" than *Dream of You*, so let me resort to an analogy. In 1977, I attended a reading by the poet Robert Creeley, who, after finishing "The Rain" (one of several poems he read from *For Love*, his first major collection, which was published in 1962) looked up from the page with a blissful smile and repeated the final three words, giving each of them equal weight. "A *decent happiness*," he said, more to himself, I thought, than to the audience. "Not as in 'are you *decent?*'" I was sitting next to one of my former English professors, who nudged me and whispered, "I bet everybody here knew he meant decent *enough* even before he explained it." I shushed him, not for fear his voice would carry, but because I wanted in on Creeley's delight in remembering a *mot juste* that he had apparently forgotten all about.

A similar elation in having gotten more right the first time than memory might allow pervades *Collaboration* and makes it a joy to listen to. So my preference for *Collaboration* boils down to something as subjective as that— and something as obvious as increased confidence on Merrill's part. Merrill has always had a slight catch in her voice, which (as *Dream of You*'s uncredited annotator observed) causes some of her notes to emerge a split second late, an encumbrance that she gradually learned to exploit to rhythmic advantage. In place of the coy flirtatiousness that was probably forced on her in 1956, she now projects a warm, matter-of-fact sensuality (over walking bass on the teasing intro of the title track, for example). A critic who was blissfully unaware that Merrill's parents were Yugoslavian immigrants once characterized her as the quintessential WASP jazz singer. Although factually incorrect, he was onto something deeper. Merrill, who was raised as a Seventh-day Adventist, forbidden to frequent nightclubs, much less perform in them, still sometimes gives the impression of being from a more respectable part of town than her material. (Though it would be unfair to make too much of it, it's worth passing along a story she told me about touring the South with an integrated package show in the 1950s: outraged at the difficulty that the black musicians faced in making accommodations, she checked in with them at a black hotel, but immediately felt out of place and beat a quick retreat to the whites-only hotel she'd originally spurned.) At least Merrill knows who she is. She smolders, but never tries to sham you—or herself—into believing that she's going up in flames, the way histrionic singers do. The wisdom of Evans's arrangements lies in his recognizing Merrill's restraint as a sign of integrity rather than self-evasion. She's never sounded better than she does here, and if she has him to thank for that, she's already repaid the debt twice over by making *Dream of You* and *Collaboration* possible in the first place.

(*August, 1988*)

THE MYSTERY
OF HERBIE NICHOLS

In 1955, two offbeat pianists found champions at jazz record companies. One was Thelonious Monk, then a cult figure whose first two Riverside albums were collections of Duke Ellington favorites and pop standards calculated to dispel his forbidding image. Although the strategy was unsuccessful—producer Orrin Keepnews underestimated Monk's gift for reshaping any tune at hand into his own creation—Monk soon gained a sizable following willing to accept him on his own terms. Herbie Nichols wasn't as lucky. The man some consider Monk's equal as both a pianist and composer was a more willing interview subject than the intransigent Monk (Nichols wrote articles on music himself, for *Metronome* and other publications): this alone should have made him a better bet for stardom. But it didn't happen that way, and why it didn't is one of the most frustrating mysteries of jazz. Nichols recorded thirty titles (twenty-nine of his own compositions plus George Gershwin's "Mine") at five different sessions for Blue Note between May, 1955, and the following April, with Al McKibbon or Teddy Kotick on bass, and Art Blakey or Max Roach on drums. Blue Note's Alfred Lion released twenty-two of these performances—among the most imaginative ever recorded by a pianist in a conventional trio setting—on two ten-inch LPs and one twelve-inch, but forgot about the remaining eight after it became obvious that Nichols wasn't going to sell on any configuration. It's tempting, but ultimately pointless, to speculate how easily Monk's and Nichols' fates might have been reversed if Keepnews had shown less perseverance and Lion had shown more.

Among those in the know, Nichols is often invoked, along with Elmo Hope and Sonny Clark, as one of the black pianists martyred by the indifference of critics, record companies, and club owners in the 1950s. But as heroin addicts, Hope and Clark contributed to their own martyrdom, unlike Nichols, who reportedly didn't even drink to excess. Though Hope and Clark were fine pianists and composers, Nichols was their superior. Yet they at least enjoyed the respect of the leading musicians of their generation—witness the many records they made as sidemen with the likes of John Coltrane, Dexter Gordon, and Sonny Rollins. By comparison, Nichols's credentials, taken at face value, were those of a blue-collar musician. Shunned by his equals, he earned his living backing gutbucket saxophonists like Hal "Cornbread" Singer, Floyd "Horsecollar" Williams, and Big Nick Nicholas in Harlem, and playing dixieland with semipros in Greenwich Village. In 1956, Billie Holiday gave Nichols what should have been his big break, when she put lyrics to his "Serenade," and retitled it "Lady Sings the Blues" (also the name of her autobiography, and her signature tune until her death three years later). But even this gained Nichols little notice. By the early sixties, he was reduced to a six-nights-a-week gig at the Page Three in Greenwich Village, backing three singers, a comedian, a shake dancer, and a stripper who began her act disguised as a man. Granted, one of the singers was the superb Sheila Jordan; and in addition to slumming celebrities like Tennessee Williams, Shelly Winters, and Lenny Bruce, the audience frequently included Cecil Taylor, Steve Lacy, Archie Shepp, and Roswell Rudd—key figures in the emerging jazz avant-garde, all of whom credited Nichols as an influence, but none of whom was yet in a position to do him much good.

Nichols died of leukemia in 1963 (ironically, the year that Monk made the cover of *Time*), at the age of forty-four. Two things have kept his name alive: the chapter on him in A. B. Spellman's *Four Lives in the Bebop Business*, a 1966 book used as a text in black studies as well as jazz courses; and today's mania for 1950s jazz, which makes reclamation projects like *The Complete Blue Note Recordings of Herbie Nichols* (Mosaic MR5-118) economically feasible. This five-record box set contains all thirty of Nichols's Blue Note performances, including the "missing" eight tracks (two of which first surfaced on a Japanese anthology released a few years ago), and eighteen alternate takes. The neglect that Nichols suffered in his own era works to his advantage now: vintage ceases to matter because the contemporary listener doesn't feel like an earlier generation got there first. Like Duke Ellington, Nichols was a master of the musical vignette—the out-of-step chorus girls on Nichols's "Dance Line" would be right at home in Ellington's *Harlem*. But because Nichols worked on a much smaller scale than Ellington, a better comparison is to the Charles Ives who wrote such short piano pieces as "In the Inn" and "Some Southpaw Pitching," especially given the dissonance, polyrhythmic complexity, and pianistic focus of Nichols's writing.

In "House Party Starting," for example, a frolicsome riff made to seem vaguely apprehensive through repetition eerily conveys the discomfort of being the first arrival at a party. In "The Spinning Song," a wistful, Asian-sounding scale that gradually turns harsh and staccato leaves no doubt that what's being spun is the wheel of fortune, and that it's always landing on bankrupt. The most provocative of the previously unissued performances is the aptly named "Sunday Stroll." But Nichols's masterpiece is "The Gig," a riot of clashing notes and tone clusters depicting a pickup band at odds with itself about what to play—sixty-seven oddball measures long with a fitful nine-bar opening strain that builds terrific momentum for the whiplash improvisation that follows. With its flatted fifths and echoes of "Tiger Rag," this is a "Strike Up the Band" conceived in the aftermath of the squabbles between the boppers and the figs, undoubtedly owing something to Nichols's nights of toil as the lone modernist in dixie joints. (Nichols believed in happy endings, so the jammers ultimately reach accord; just as a bevy of happy, talkative guests finally shows up in "House Party Starting.") It's significant that "The Gig" is from a session with Roach and McKibbon, a combination more in empathy than Blakey and Kotick were with Nichols's desire to tap the hidden melodic potential of walking bass lines and snare drum overtones. Nichols wanted bass and especially drums to provide melodic coloration as well as propulsion; or, as he put it, he wanted rhythm to *sound*, as it does in African music (to which he was one of the first American jazz musicians to pay serious attention). Nichols's tracks with McKibbon and Roach rival Ahmad Jamal's with Israel Crosby and Vernel Fournier as the best-integrated piano-bass-and-drums performances of the 1950s, and they enjoy a big advantage over Jamal's in not being as coy or overformatted. When he wasn't experimenting with the kind of extended form represented by "The Gig," Nichols was subverting thirty-two-bar convention from within, with between-the-bar-line chromatics and arpeggios, and teasing intros and codas that were all part of the grand design but didn't count against the final tally (the best example of this is in "Steps Tempest"). As an improviser, he was strictly a theme-and-variations man, as he could well afford to be, because his compositions prefigured practically every conceivable harmonic option. (For this reason, the Mosaic box's alternate takes generally differ from the masters only in tempo. They are a justifiable luxury, however, because with the exception of a handful of sides reissued on Savoy, a 1957 Bethlehem LP, and incongruous dates as a sideman with swing trumpeters Rex Stewart and Joe Thomas, this is all the Nichols there is on record.) Nichols was like Monk in this and in many other ways, including the vestigial echoes of Harlem stride in his left hand and his uncanny ability to conjure the secret notes between the keys. In 1943, as a regular columnist for the Harlem-based periodical *The Music Dial*, Nichols had the distinction of giving Monk what was possibly his first review. After praising Monk's "rhythmical melodies," Nichols chastized him for his partiality to "certain limited harmonics" that would keep him from taking a place beside Teddy Wilson and Art Tatum. As a judgment of Monk, this was absurd. But as a statement of Nichols's own values, it was perfect: Tatumesque flourishes of the sort that Monk disdained prevented Nichols from sounding like Monk's twin brother.

In confirming Nichols's greatness, *The Complete Blue Note Recordings* leaves hanging the question of why his peers turned deaf ears to him. To say that his music was "difficult" won't suffice. So was Monk's, after all. As a result of Nichols's diverse experience, he had to be a skilled and versatile accompanist (in a liner note for an earlier reissue, Roswell Rudd likened playing with him at a sixties loft session to being "a soloist in a grand concerto, with a multitude of other voices leading, supporting, and responding"). When his own records didn't sell enough to justify keeping him under contract, why didn't Blue Note find use for him as a sideman, or commission pieces from him for others on the label's roster? Were Ike Quebec, Lou Donaldson, and Stanley Turrentine so different from the saxophonists Nichols worked with in Harlem? Lion deserves the credit he's received for recording Nichols, but he also deserves censure for not sticking with him, as the less-deified Keepnews did with Monk. In a sense, Nichols's Blue Note recordings foster what may be a misleading impression of him as a composer of flawless miniatures for piano, bass, and drums. For an inkling of what he might have sounded like with horns, you have to turn to two eighties albums featuring the Dutch pianist Misha Mengelberg: *Regeneration* (Soul Note SN-1054) with Rudd and Steve Lacy, and *Change of Season* (Soul

Note SN-1104) with Lacy and George Lewis. Toward the end of his life, Nichols told Spellman that too few black instrumentalists knew what to make of musical notation: the lament of a composer who knew that his un-recorded works would survive only on score paper—which meant that they might as well never have been written.

I asked a few of Nichols's musical associates for their recollections of him, and if they had any explanation for why he was so unjustly overlooked.

Big Nick Nicholas: "When we played together in Harlem, we'd play his tunes and mine, plus standards we both liked. He was a good accompanist who played his own style no matter who he was playing with. You'd always know it was him. But he was a very dedicated musician who was willing to play any kind of music in order to make a living. He played dixieland because it was steady work. The alternative would have been to work a week every once in a while, then not work again for six months. You can't live like that. He wasn't an opportunist. There are two kinds of musicians. There are guys like Herbie Nichols and myself who don't push, and there are the guys who are out there pushing every day, which is unfortunately what does it in America. Some guys who aren't very good are very good at telling you how good they are. He wasn't one of those."

Sheila Jordan: "Jackie Howell, the owner of the Page 3, adored Herbie. He always had an attentive audience there—she made sure of that. I still have a lot of the chord changes he wrote out for me: 'Lush Life,' 'Love for Sale,' 'Lady Sings the Blues,' and a song of his that he never recorded called 'My Psychiatrist.' Most of the singers at the Page 3 didn't have their music together. They'd ask him for 'I'm Confessin'' in D-flat, but he'd never complain. He was a very good-looking man, very tall and mysterious, and always well dressed. I remember one time he was standing outside of the club and he saw me getting out of a car driven by a man who needed a shave. He asked me what I was doing with a gangster. I said 'Herbie! That's the drummer in the trio that's here on Mondays, your night off.' He couldn't imagine a musician not being well groomed. He'd stand in the background when he wasn't playing, never saying very much, but you knew he was taking everything in—he saw the guys coming in, buying the strippers drinks, and he feared the worst about me when he saw me getting out of that car. He was such a sweet man. You know how musicians on the scene can be, always telling each other 'I have a record date coming up, and you can be on it if I can be on yours.' Herbie wasn't like that. He was very reserved—not unfriendly, but very dignified and very shy."

Archie Shepp: "When I knew him, he knew he was ill, and he was starting to become rather disillusioned. He felt that there was too much politicking in the music. He was a genius, but he got lost in the shuffle, which seems to be an un-fortunate aspect of this system of ours—to lose geniuses in the shuffle. I don't know why he didn't get more work as a sideman. He could play with anybody, but I don't think he would have been comfortable as a sideman in a conventional jazz setting, any more than Monk would have been. That's a dilemma inherent in our music: should a gifted soloist and composer like Herbie be forced to bend in order to fit into a traditional sideman's role?"

The Herbie Nichols who emerges from these recollections was a quiet man disinclined to tout his own virtues, which is evidently what it took to get ahead in the competitive jazz scene of the fifties. But there must have been more to it than that. Was Nichols gay, or was something else that no one wants to talk about an alienating factor? You could argue that his two-handed attack struck his contemporaries as old-fashioned in the wake of Bud Powell. But what about Erroll Garner and Oscar Peterson? Did the period's more strung-out musicians resent Nichols for being such a straight arrow? Maybe so, but didn't they adore Clifford Brown for just that reason? If Nichols had survived, would endorsements from Rudd, Shepp, Cecil Taylor, and others associated with free jazz have won him more attention? It's debatable. Nichols complained to Spellman that a musician had to be a junkie or an Uncle Tom in order to be accepted by audiences; my guess is that he would have considered the militancy that became *de rigueur* in the wake of free jazz yet another loathesome black stereotype. Besides, with the obvious exception of Taylor, there was little call for pianists in the new music. As Shepp points out: "It isn't a question of what would have happened if Herbie Nichols had lived longer. It's a question of whether circumstances would have been any better for him than those he *had* lived under. If he had gone on living under the same circumstances, he might have become an extremely bitter man."

Nichols is going to remain a mystery. The twenty-four-page booklet that comes with the Mosaic box gives us as complete a picture of him as we're ever going to have. It includes Rudd's track-by-track analysis, testimonials by Rudd and others, a Nichols discography, samples of his poetry and prose (not his Monk review, though), and Francis Wolff's phantomlike snapshots of him—all this, plus his music, which is so magnificent you can't help wondering why he wasn't allowed to make more.

(*April, 1988*)

THE CANTOS
OF CECIL TAYLOR

I *think* that the second time I heard Cecil Taylor in concert was in a church. If so, it was a proper setting, because his music demands the most obsessive sort of devotion. I know for sure that the first time I saw him was at a New Jersey college in 1973. The auditorium was so crowded that people were being seated within inches of him and his sidemen, Jimmy Lyons and Andrew Cyrille. From my vantage point directly opposite the piano, Taylor's fingers were a blur. I was close enough to be hit by his sweat, and to ascertain that the force of his attack really did move the piano back on its wheels (I probably only imagined that it levitated once or twice). For the first twenty minutes or so, the faculty members in the front rows watched him in bewilderment; then in anger when they realized that a polite retreat to the exits between numbers was out of the question, because this was going to be a nonstop performance. Waiving decorum, they begin to flee anyway; minutes after the first departure, the hall was almost empty. This was after the deaths of John Coltrane and Albert Ayler, when the jazz avant-garde was pulling in its horns, courting favor with lulling ostinatos and hymns to the creator (its own version of silly love songs). I admit taking satisfaction that someone was still chasing the unconverted from the room with ironfisted music. By this point, the unconverted know enough to stay away, and Taylor has acquired enough of a cult following to sustain his career without them. But I often suspect that those who adore Taylor (myself included) and those who dread him have more in common than they know. Both factions perceive his performances as uninterrupted spasms; the difference is that uninterrupted spasms are exactly what some of his fans are after. I know I was that night in New Jersey, but I'm older now.

   Did Whitney Balliett once liken Taylor to a blacksmith rolling up his sleeves and hammering out anvil choruses?* The manly smelter of Taylor's music, its jackhammer speed and mythopoeic energy—these are celebrated in every description of him at work. But what I like about the blacksmith image is that it also recognizes symmetry, purpose, craft. The tension of hearing Taylor

---

*What Balliett actually wrote was, "He is a hammer and the keyboard is an anvil."

balance discipline and abandon for over an hour can be almost sexually exhilarating, but it would be a mistake to praise him solely for his ability to keep it up (as it were). I used to wonder how many listeners possessed the celerity and stamina to go the distance with Taylor, and now that he's acquired a sizeable following, I wonder if some of his loyalists—and he himself, at times— overvalue the orgiastic, purely physical aspects of his art. (To his credit, Taylor objects to the violent imagery sometimes used by whites to describe his performances, and black music in general. Or so I am told by a colleague who incurred Taylor's wrath by writing—in praise—that hearing him in concert was like being mugged.)

The length of Taylor's performances (one piece can consume an entire two-hour concert) creates needless difficulties, which records only exacerbate. The difficulty with most visionary music is in finding an easy way in. With Taylor's, it's in finding a graceful way out. His concerts have no clear beginning or ending. His stage entrances are protracted by his introductory chanting, which irks even some of his longtime admirers, but intrigues me as a public unburdening of the vocal coaching method he's said to employ at rehearsals as an alternative to notation. (Is it also his nod to showmanship?) His exits are swift: he often sneaks off without an encore when the piano provided for him hasn't met his standards. But however much one's attention may wander while hearing Taylor in concert, at least there's the visual reference point of watching flesh kneading and pummelling ivory, and the feeling afterwards that you, Taylor, his sideman, and the rest of those assembled have all *been* through something together. On record, the length is a built-in alienation factor: the spell is broken whenever the phone or doorbell rings.

What can make listening to Taylor on record even more frustrating is that few of his releases over the past two decades have been conceived as recordings in the first place. (I admit to preferring his studio albums, his solo recitals, and his pre-seventies work in general: those albums like *Into the Hot*, *Unit Structures*, and *Conquistador*, from the period when his accompaniment was crowded and orchestral, but his improvised solos were relatively spare—before his music overheated and its containers cracked.) Most of Taylor's albums over the last twenty years have documented concert performances, and *Live in Bologna* (Leo LR-404/405), a two-record set preserving one eighty-minute piece and a ten-minute encore, is no exception. Like all of Taylor's records, this one clangs with greatness, but I despair to speculate when I'll again find an hour and a half to give it the undivided attention it asks for and deserves. Is there time in one life for family, career, *The Mahabharata*, and Cecil Taylor? Should art make such unreasonable demands?

Live in Bologna introduces the newest edition of the Cecil Taylor Unit, with

Carlos Ward given the unenviable task of replacing the late Jimmy Lyons, the alto saxophonist who played a vital but largely misunderstood role in Taylor's groups for twenty-five years. Lyons anchored Taylor in jazz tradition, but some listeners initially dismissed him as a bopper out of his depth after hearing him on the 1962 album *Into the Hot*. In observing tonal centers where none existed, Lyons was no doubt complying with Taylor's wishes. On the half-dozen pianoless albums that Lyons made under his own name, his needle-sharp pitch and untethered harmonics revealed a kinship to Ornette Coleman, and provided an unexpected answer to the question of what Coleman and Taylor would have sounded like together: the combination was unimaginable, because Coleman would never have bent to Taylor's will, as Lyons did so unselfishly.

Understandably, Ward sounds less certain of what Taylor expects from him. The hymnlike tone that he displays as a member of Abdullah Ibrahim's group, Ekaya, is thwarted here, and he lies low much of the time. Toward the end of side three, however, he asserts himself with a chantlike four-note phrase similar to Coltrane's *A Love Supreme*, and the ensemble responds to him at fever pitch. And although I could live without flutes in general (unless we're talking Eric Dolphy), Ward's use of this horn, together with the drummer Thurman Barker's marimba, brings new and immensely attractive shadings to Taylor's music. Much of these four sides is given over to a shimmering chamber improvisation with echoes of Balinese gamelan music—but without the solipsism that defeats most such attempts at cultural cross-breeding. In a sense, *Live in Bologna* represents Taylor's consolidation of recent trends elsewhere within the jazz avant-garde that he helped to father, although only time will tell whether this is just a transitory response to new personnel. (The violinist Leroy Jenkins's combination of street smarts and classical intonation makes him the Unit member most in tune with Taylor; the bassist William Parker, though only intermittently audible, provides a solid harmonic bottom.) There are no piano episodes here as lengthy or glancing as those on *Spring of Two Blue-J's* and *Dark to Themselves*, but Taylor's leadership of this group is unyielding. Every note, regardless of which instrument it's actually played by, seems to pass through his keyboard first. The volume is generally softer here than on most of Taylor's group albums, and this may help to convince even the most skeptical of his critics that his music is operatic in its melodic underpinnings, as well as in its tumult and scale.

On *Chinampas* (Leo LR-103), a companion release, Taylor recites six of his poems, accompanying himself on various percussion instruments, but not playing piano. This is a limited pressing of five hundred copies, strictly for

Taylor completists. Although Taylor's poetry resembles his music in its insistent repetition (call-and-response is his favorite method of development, even when playing solo piano), it's otherwise difficult to judge, because he neglects to include a printed text (perverse of him, in light of the fact that he printed his poetry on the sleeves of so many albums without recitation). At times it's difficult to comprehend what he's saying, and his meaning depends more on sound than on sense ("It be crystal/It be crystal/crystallized./It be crystal/crystallized smashpoint./It be crystallized/It be crystallized smashpoint of no/joint./Covenant stone/Covenant stones/are exposed and/consumed articulation.") One poem beginning with overdubbed parallel texts is a vendetta against Stanley Crouch, the jazz critic who violated Taylor's privacy by revealing that he was gay; it joins Ezra Pound's diatribes against the Rothschilds in proving that poetry isn't necessarily the best revenge.

I suspect that Pound's *Cantos* are among Taylor's poetic models, along with such Pound-haunted epics as Louis Zukofsky's "A" and Charles Olson's *The Maximus Poems*. At the very least, Taylor echoes the irascible, wise-fool delivery Pound used on the albums he recorded for Caedmon while at St. Elizabeth's. Taylor's invocations of Damballah and bursts of solfeggio serve much the same purpose as Pound's pronouncements on usury, untranslated Cavalcanti, and untransliterated Confucius: the reinvention of poetry as the movement of thought through the mind of an erudite man. Pound's *Cantos* fostered the idea (now shared by most poets working in larger forms) that the individual poem is an installment in an ongoing life's work. In this sense, Taylor's *Cantos* are his performances with his Unit, which make as many unreasonable demands as Pound's poetry, and are arguably no less worth the effort. What, then, is *Chinampas?* Taylor's equivalent of Pound's opera *The Testament of François Villon*—a great artist's eminent domain.

(May, 1988)

THE HOME
AND THE WORLD

"Every individual in South Africa is traumatized by apartheid—the oppressors perhaps more so than the victims," said Abudullah Ibrahim, the South African pianist and composer formerly known as Dollar Brand. "At least *we* have hope for the just society to come: a South Africa that belongs to all of its peoples, regardless of color, as stated in the Freedom Charter. The oppressors' only hope is to forestall the inevitable. The only wisdom they can teach their children is to always carry a gun."

I visited Ibrahim and his wife, the singer Sathima Bea Benjamin, on June 12, 1986, the morning that South African president P. W. Botha declared a nationwide state of emergency in anticipation of the tenth anniversary of the Soweto uprising. Kneeling in front of a video monitor, Ibrahim—wearing a dashiki and corduroy jeans—tried without success to tune in C-SPAN for its live coverage of Botha's address to South Africa's parliament. "The police have rounded up all the opposition leaders," he told me in a melting-pot patois that was British in formality but African in emphasis and lilt. "What is the logic of that—to remove from circulation the only people who can say to the masses, 'Hold back'? The confrontation is near. It is just a matter of time. It will be this weekend."

Tall and fit, with long, graceful, active hands and the bearing of a diplomat (which, in a sense, he is), Ibrahim was born in Cape Town in 1954, but has spent most of his adult life in voluntary exile in Europe and the United States. He is a cultural eclectic—no different from most global wanderers in that respect, though his being born a member of a colonialized people undoubtedly gave him a running start. (Moreover, his mixed racial heritage, which classifies him as "colored" rather than black by South African law, puts him in an assimilationist position remarkably similar to that of the New Orleans Creoles and mulattoes who played a decisive role in the gestation of jazz.) But like most political exiles, Ibrahim is also a displaced cultural nationalist whose longings for home filter his perception of new surroundings. The South African sound he once eloquently described as being a synthesis of "the carnival music heard every year in Cape Town, the traditional 'colored music, the Malayan strains, and the rural lament" remains the thread that keeps his music from seeming crazy-quilt despite its patchwork of cross-cultural borrowings. Those include West African ceremonial and popular rhythms, Moslem incantation, British military-band concord, gospel sanctimony and minstrel sanguinity, French impressionism and modal reverie, Monkian dissonance, and Ellingtonian Cotton Club panache (what's surprising about Ibrahim's music, given its origin in a troubled place, is its defiant joy). A mesmerizing pianist whose rhythms splash across the keyboard in thunderous ostinato waves, Ibrahim is "beyond category," to borrow a pet phrase from Duke Ellington, one of his earliest admirers (and the producer of his first American LP, in 1962).

Over the last twenty years, Ibrahim's solo piano recitals have evolved into lengthy, continuous medleys of his own attractive themes and those of favored composers like Ellington, Thelonious Monk, Billy Strayhorn, and Eubie Blake. The aspect of his work is best documented on *Autobiography* (Plain-isphare PL-1267 6/7), a two-record set recorded live in Switzerland in 1979. In his otherwise perceptive liner notes for *Autobiography*, the musicologist Wilfrid Mellers likens Ibrahim's medleys to collage—a comparison better suited to Keith Jarrett than to Ibrahim, because what distinguishes Ibrahim from ramblers like Jarrett is his dedication to composition and his insistence on full thematic exposition. There is nothing fragmentary or episodic about Ibrahim's treatment of the myriad themes on *Autobiography*, or about the way he builds from rhapsodic, harplike glissandi to house-shaking honky-tonk, or from thick, frigid clusters to long-steepled, liturgical-sounding tremolos. (Exactly how many themes there are on *Autobiography* is difficult to say. The jacket lists eight Ibrahim originals, plus Monk's "Coming on the Hudson," Strayhorn's "Take the 'A' Train," and the standard "I Surrender, Dear." But Blake's "Memories of You" frames the homages to Monk and Strayhorn, and I hear at least five other unidentified Ibrahim pieces, in addition to Charlie Parker's "Ornithology" and Monk's "Four in One.")

Ibrahim's piano solos blueprint larger designs; as with Ellington, you can hear the colors of an orchestra bleeding together and separating in the voicing of each chord. But with the exception of *African Marketplace*, an out-of-print masterpiece from 1980 (Elektra 6E 252), Ibrahim's large group recordings haven't represented him very well. His music is holistic and horizontal, the melodies rolling steadily over and around embedded rocks of rhythm. Merely harmonizing the heads of his tunes and running their changes (as many American musicians have done), you wind up on dry, flat land or in murky swamp. Ibrahim's approach to improvisation, which arises out of a folk tradition that uses improvisation to extend melody and rhythm, puts him at

odds with most American and European jazz musicians, for whom improvisation is an end in itself. "Improvisation is not standing up there and playing
twenty choruses," he once told the critic Don Palmer. "It is using what's at
hand."

With Ekaya—the fluid septet that Ibrahim formed in 1983 and with which
he has recorded two albums for his and Benjamin's Ekapa label—Ibrahim
found sidemen confident enough to shun chordal rhetoric and crest on his
rhythms. Even though the band is a septet, it's tempting to think of its two
albums—*Ekaya* (*Home*) (Ekapa/BlackHawk BKH 50205-1), and *Water from
an Ancient Well* (Ekapa/BlackHawk BKH-50202-1)—as big-band records,
because Ibrahim's voicings for trombone and three saxophones are so plump
and varied in texture, and because he showcases his soloists so strategically.
With Ibrahim's rhythms brewing below him, the young tenor saxophonist
Ricky Ford reverts to his Sonny Rollins influence, and he's never sounded so
passionate, fierce, or suave. But Ibrahim's most simpatico accomplice is the
Panamanian alto saxophonist and flutist Carlos Ward, whose piercing, jubilant
leads define the band's ensemble sound, and whose solos match Ford's in
crackle and invention. Perhaps the truest proof of Ibrahim's genius as a
bandleader, though, is his ability to coax trim, inspired improvisations from the
trombonist Dick Griffin and the baritone saxophonist Charles Davis, two
journeymen who are usually more prolix.

Ekaya represents a compromise between the large troupe Ibrahim desires and
the small group that economics dictates. "The ideal situation would be to have
as many people as possible," he told me. "The more people the better! Not just
instrumentalists, but singers, painters, sculptors, poets, nutritionists, physicians, and martial artists, to show that music can provide a conducive
atmosphere for all other daily activities, as it does in more traditional societies."

He likens his empathy with Ward to that which he once enjoyed with the late
Kippy Moeketsi, "the Charlie Parker of South Africa," in Ibrahim's estimation,
and the father figure in the Jazz Epistles, the Cape Town combo that begat
Ibrahim and the trumpeter Hugh Masekela in the late 1950s. "Kippy was the
first to insist that we recognize the wealth of musical influences available to us
as South Africans and not look exclusively to the U.S. for inspiration. That was
at a time when our sense of ourselves as a nation was being born, and Kippy's
musical philosophy was a part of that. He was a pillar of strength against those
who would have had us believe that we were inherently inferior as musicians."

Ibrahim remembers the seaport of Cape Town as "a cosmopolitan mixture of
Xhosa, Zulu, British, Dutch, Khosian, and Malayan traditions. As musicians
there, we heard everything, and we played everything. There was even a Cape

Town Symphony Orchestra and an opera company that performed *Madame Butterfly, La Traviata, La Bohème*. All of my early piano teachers were well versed in the European classical tradition. American merchant vessels would dock in the harbor, and we would run to meet these ships, because the African-American sailors on board would have American jazz records to sell to us. It was from these sailors that I received the nickname 'Dollar' Brand—my real name was Adolf. My grandmother was a founding member of the African Methodist Episcopalian church in South Africa, and my mother was the church pianist, so I was exposed to the black American tradition through the so-called Negro spirituals I sang in the choir.

"I was familiar with Islamic customs long before embracing Allah, because the Christian and Moslem communities were very close in Cape Town as a result of so much intermarriage. The Moslems would know when we were celebrating Christmas or Easter, and we would know when they were celebrating Ramadan. Many families observed all the religious holidays across the board. Ethic diversity—that's the way it is all over the world. There is no pure race. Yet in South Africa, the races that have mingled since the beginning of time are separated by law. They have this multitiered system, which is actually a class system. On top, you have the whites. The second stratum is the so-called 'colored,' the Asians, and the Indians. And at the bottom, the blacks, who are broken down further by tribe."

Ibrahim began to read from a newspaper clipping. "'More than a thousand people officially changed color in South Africa last year. They were officially changed from one race to another by the government. Details of what is called The Chameleon Dance were given in Parliament: 715 coloreds turned white, 19 whites became colored, one Indian became white, 15 Indians became colored, 43 colored became Indian, 31 Indians became Malayan, 15 Malayans became Indians, 349 blacks became colored, 20 coloreds became black, 11 coloreds became Chinese, 3 coloreds became Malayian, 3 Chinese became colored, and 3 blacks were reclassified Malayan.

"'No whites became black, and no blacks became white.'"

Sathima Bea Benjamin, who was born in Cape Town in 1936, remembers it as "a beauteous place, with mountains, sea, perfect climate, flora and fauna, exotic birds, and gorgeous sunrises and sunsets. There seemed to be music in the air, always. Is that possible, or is it just the way that I remember it?"

Benjamin and Ibrahim asked to be interviewed separately. "I am Sathima, and he is Abdullah, and although we love each other very much, we have separate careers." Whereas her husband tends to express himself in parables, Benjamin speaks in the down-to-earth fashion of a woman who has spent a good deal of her life preparing meals and caring for children. "I don't perform as often as I would like, because I cannot. I have three roles, the most important of which is running the house and looking after our children [son Tsawke, fifteen, and daughter Tsidi, ten]" she said during a conversation she crammed in between picking Tsidi up from school and keeping an appointment with a tax lawyer. "Second, there is my music, and third, Ekapa, the record company. My life is simpler now that we've turned over Ekapa's distribution to Black-Hawk Records. I spend much less time at the post office. I guess I actually have four jobs, because Abdullah and I act as our own agents. When he's away, which he is much of the time, his phone still rings, and someone has to answer it.

"I was raised to be a dutiful sort of person. My parents were separated by the time I was five, and I was raised by my paternal grandmother, who was very strict, very proper, very British in her ways, although she was quite African-looking. I was a lonely child, and along with my daydreaming, which I indulged in constantly, music was my only solace. I listened to American singers on the radio: Ella Fitzgerald, Nat 'King' Cole, Joni James, Doris Day. I was very attracted to the music of Victor Herbert, songs like 'Indian Summer' and 'Ah! Sweet Mystery of Life,' which I still perform. Musicians ask me, 'How do you know those songs? You weren't around in the 1920s.' And I tell them, 'No, but my grandmother was.' When I joined the choir in high school, I noticed that the director never assigned me solo parts, even though I had a very strong voice. I asked him why, and he said, 'Because you sweep. You slide up and down the note, instead of staying directly on it.' That meant nothing to me at the time, but, in retrospect, it shows that I was unconsciously trying to imitate the black American singers I heard on the radio.

"After completing college, I taught school. But on the weekends, I sang in the nightclubs in the white areas, where black and so-called 'colored' enter-tainers were allowed to perform but were not allowed to mix with the customers. We had to sit in the kitchen during intermissions, just as black musicians were having to do in the American South. There wasn't such a strong ban on U.S. literature then, so I was able to read a good deal about black Americans, and I felt a bond with them, with their longing to be free." Lighter in complexion than her husband, she, too, was classified as colored. "Some-times people ask me, 'Oh, Sathima, why do you call yourself black? I have to be careful not to overreact, but *inside* I overreact. Black is not a color, it's an experience. And in South Africa, there are only two possible experiences. I was never privileged to know what the white one was. That makes me black."

She and Ibrahim first left South Africa to live in Zurich in 1962, and it was there that they encountered Ellington, "the first American that either of us had ever met—and thank God it was him. Abdullah and I had been brought together by our mutual love for his music, so to actually meet him was like a fairy tale. I still marvel at how truly *grand* he was." In addition to producing *Duke Ellington Presents the Dollar Brand Trio* for Reprise, Ellington also supervised Benjamin's never-released debut. "Duke and Billy took turns accompanying me at the piano. I remember Duke standing in the control room at one point and saying to Billy, 'Can I hear some birdies?'—meaning could he play something sweeter than the wild things he was playing behind me. And Strayhorn answered, 'I am playing birdies—condors.' Fortunately, from having sung with Abdullah, I was used to heavy, dissonant chords."

Benjamin told me she was contemplating an album of Strayhorn compositions as a companion piece to her 1979 album, *Sathima Sings Ellington* (Ekapa/BlackHawk BKH-50201-1) About ten years ago, relatively late in her career, she began to write her own material. "At first, I was hesitant even to show my songs to musicians, because I know I'm not a trained composer like Abdullah. But the musicians I work with—Kenny Barron, Buster Williams, Billy Higgins—encourage me. 'Remember, Sathima, music isn't just the notes.' they tell me. 'It's the feeling, too.'" She still "sweeps": there's a catch in her voice that would be easy to mistake for coquettish affectation if not for the sob it barely holds in check.

"The reason I like living in New York is that I don't feel that different from anybody else," Benjamin said. "There are people from so many different countries here, so many different nationalities of people you encounter just riding the bus. It helps me to accommodate myself to the fact that I've been uprooted." Despite living in the Hotel Chelsea for almost a decade, Benjamin took the oath of U.S. citizenship only last year. "For most of those at the swearing-in ceremony, it was a joyous occasion. And it was for me, as well. But I went through a period of soul-searching that I can't begin to describe, because South Africa will always exert such a pull."

"I am still a citizen of South Africa," declared Ibrahim, who nonetheless said he intends to apply for U.S. citizenship eventually. "I always will be. Sathima and I are in strategic retreat, but we expect to return. We used to say that our music was our home. For years, we were free to go and return as we pleased, so long as we didn't make any overt political statements. Soweto changed all that. The struggle had reached another level, and it was important for us as artists to play a more visible role." He and Benjamin were last in South Africa in 1976, and cannot return there even to make final amends to the mother Benjamin barely knew, now seventy-five and gravely ill. If they were to risk a visit, it isn't likely that they would be permitted to leave: the Botha government, in retaliation for their fund-raising for the outlawed African National Congress and the forthright political tone of much of their recent work, last year refused to renew their passports. "Our children were born in Africa—I insisted on that much," Benjamin said. "But New York is where their memories of growing up will be, and memory is a very powerful force."

Unlike their parents, Tswake and Tsidi consider themselves New Yorkers, not South Africans. The home that Benjamin and Ibrahim knew as children no longer exists—at least, not as they remember it. "Perhaps the most notorious of all the government removals in the urban areas was the destruction of District Six, home of 30,000 people classified as 'colored' in the heart of Cape Town," writes Francis Wilson in the preface to *The Cordoned Heart*, a collection of photographs from South Africa.

> District Six was a diverse society established over a period of two hundred years. Its inhabitants prayed in mosques and churches and syna-gogues. . . . There were sounds of fruit vendors, homemade banjos, muezzin calling the faithful to prayer, the laughter of children, Hindu funerals and Muslim weddings. . . . It was a cultural centre. At one time or another, it had been home to musicians like Abdullah Ebrahim [*sic*], writers like Richard Reve, political leaders like Cissie Gool. . . .
>
> In 1966, the district was proclaimed white. The order setting in motion the removal of the citizens there and of the destruction of their homes was signed by the then-Minister of the Department of Community Development, P. W. Botha. . . . Virtually every building was broken up by bulldozers. Hundreds of strong brick houses, some of them over a century old, were reduced to rubble. The vibrant world of District Six became an empty wasteland. It was as if there had been war.

"District Six is synonymous with what happened all over South Africa as a result of the Group Area Act," Ibrahim said when I read him the passage. Behind him, I noticed a Moslem prayer calendar tacked to the living-room bulletin board, alongside more earthly reminders about piano pupils and furniture in storage. "Allah says, 'Fight injustice wherever you find it, or you will become one of the unjust.' I am homesick for South Africa, but not for the home that is there. Even people still living in the houses they were born in in South Africa are homesick for another home, the spiritual home. Allah says, 'I do not burden a soul with more than it can bear. Those who leave their homes for my sake, I will provide for them a better home.'"

POSTSCRIPT

In August of 1986, the South African embassy surprised Benjamin by granting her request for a one-time-only entrance-and-exit visa on compassionate grounds, to visit her mother. "It was the same beautiful place physically, and the morale there was something to behold. There was a determination there, a unity among the diverse tribes that I had not seen before, that convinces me it is just a matter of time," said Benjamin, who turned down requests to sing even informally during her two-week visit, not wanting to violate the ANC boycott she had besought other performers to honor.

A SCHOOL OF ONE

The Algerian-born pianist-turned-drummer Errol Parker and I were sitting in his fourth-floor walk-up on First Avenue, talking about the multiple, juxtaposed, and extended rhythms that underpin his music. These rhythms were North African in conception, Parker said, sliding behind his traps for a demonstration.

"You see, in Africa, they have all sorts of drums, but they do not have traps," he said, tapping a pattern with drumsticks on the conga mounted to his set where the snare normally sits, then gradually compounding the rhythm on high hat, ride cymbal, bass drum, and three sets of tom-toms until he had to shout to be heard. "So what I am doing, basically, is adapting hand drumming techniques to the trap set, playing each drum for its sound, like in African drumming. This is what all jazz drummers do during their solos. But when they accompany, everything revolves around the snare, with the other drums used only for coloration and punctuation. In my music, as in African music, no matter what instrument is up front, the drums are the dominant voice at all times.

"I am a terrific drummer," he marveled after concluding the impromptu work-shop. "I can really swing a band."

What needs to follow is a description of Parker proving as much on a bandstand, compensating for his lack of drummer's coordination with grim determination. (With sticks in his hands, Parker looks like Alley Oop—it's got something to do with his outgrown Prince Valiant, his sleeveless African shirts, his perpetual look of glum determination. He doesn't look agile enough to solo or trade fours, but his music ignites without such tinder.) Such a description requires a long memory or a vivid imagination. Despite a lineup that has at various times included the alto saxophonist Steve Coleman, the tenor saxophonist Bill Saxton, the trombonist Robin Eubanks, and the trumpeters Wallace Roney and Graham Haynes—young musicians with far greater name-recognition than their middle-aged leader—the last job the Errol Parker Tentet played was a one-nighter at S.O.B.'s in New York more than a year and a half ago. The group hasn't even rehearsed at full strength since then, although Parker has worked as a pianist in a futile attempt to drum up interest in himself as a bandleader. "As a solo pianist, I can steal the show from anyone," he says. "That plus the Tentet makes me a double attraction, and this will pay off eventually."

As you may have surmised, this pianist leader of a pianoless tentet is an underdog with no shortage of self-esteem (he's approached both Milos Forman and Louis Malle with an autobiographical script entitled A *Flat Tire on My Ass*). African, though not black (and thus a nonentity to world-beat trendies—curiously, Parker is himself predisposed against white musicians), Parker was born Raph Schercroun in Oran, a seaport town in the French colony of Northwest Algeria, in 1930, the son of French-speaking middle-class parents who resorted to Arabic only when they wanted to keep their conversation secret from him. As a child, Parker showed a "disposition for sculpture, always molding clay or anything else I could get my hands on." At seventeen, he was accepted at the École des Beaux-Arts, a famous Paris art school. He was a self-taught pianist, despite the efforts of his mother (a good sight-reader with an affection for the French impressionists) to teach him proper keyboard technique: "She tried to teach me, but it became apparent that I wasn't made for classical music. Sight-reading was of no use to me since the music I liked was only on records, not sheet music. Even now that I am an experienced arranger and have played in big bands, I am still not a good sight-reader."

The first jazz record to make an impression on Parker, when he was fourteen, was "I Know That You Know" by Rex Stewart and his Footwarmers with Django Reinhardt, which the movie theater next door played during intermissions. Reinhardt was also the first musician Parker heard in Paris after enrolling at the Beaux-Arts in 1947. In 1950, Parker recorded with the Gypsy guitarist—a possible role model in being neither a white European nor a black American. "I didn't realize how much we had in common because I didn't yet realize how stubbornly African my musical instincts were until I was exposed to black American culture. For a musician, coming to New York is a form of psychoanalysis. Because of the competition, you have to mobilize all your forces and weigh your strong points against your weak points. When I came here [in 1968], my only solid ground was rhythm, and I soon learned where it came from."

But we're jumping ahead of ourselves. Feeling that Paris offered better opportunities for a sculptor, Parker considered himself only an avocational musician, despite record dates with James Moody and Kenny Clarke, a steady gig with Don Byas from 1955 to 1958, an organ record as Ralph Schecroun (a corruption of his real name), and a string of mood albums in the style of Erroll Garner (these led to Parker's permanent name change). When I asked Parker if he retained any particularly vivid memories of jazz on the Champs Élysées, he laughed and said, "Oh, so you want stories, do you?" and proceeded to tell me about the night that Stephane Grappelli serenaded Jayne Mansfield with "Exactly Like You." "She asked to see his violin, and right there at the table, she played a version of the song that outswung his." Sounds apocryphal, but Parker swears he saw it happen.

In 1964, Parker gave a one-man show of his sculpture, which he described as "intuitively African, like my music." Against the recommendation of the gallery owner, he priced his works so high that none sold. In need of fast income, he approached a French record company with "Lorre" (*Errol* spelled backwards), a Bach-*cum*-jazz novelty that was getting a big response from his audiences at the Club Saint Germain. "Lorre," which Parker now renounces (quickly changing the subject when I asked him if I could hear it), hit in Europe, and for a few months, Parker lived "like a playboy," until injuries sustained in a car accident made it impossible for him to maintain the glib style that had brought him popularity. (The damage was to his deltoid region; the deltoid is the muscle that raises the shoulder away from the body. Though Parker says he is "seventy percent recovered," the injury may account for the low action and stiff look of his drumming.)

Faced with the prospect of developing an entirely new keyboard approach, Parker decided that the jazz he'd been playing was a pale imitation of the real thing. Dissatisfied with European rhythm sections, he emigrated to New York, where he quickly found work as house pianist at La Bohème (a now-defunct club near Lincoln Center) and resumed his friendship with Dizzy Gillespie, whom he first met in Paris in the late forties. But by 1968, with jazz in eclipse, endorsements from such established musicians as Gillespie and Duke Ellington (for whom Parker, while still in Paris, recorded two tracks for a private recording intended as a gift to the ailing Billy Strayhorn) no longer opened doors, so Parker was soon reduced to taking anonymous rhythm 'n' blues jobs. On one of these, he made a chance discovery: "Under a deceptive appearance of simplicity, the blues is a bi-tonal music. One night, I started to play dissonant chord structures on top of the simple bass line I was playing along with the bassist. Nobody complained. That night, I had a glimpse of what my future band would sound like." He formed the Errol Parker Experience to play a music he has described elsewhere as "dissonant but not avant-garde, funky but not fusion, and made up of polytonal chord structures that could be shifted around freely . . . (but) anchored to a strict bass line simple enough to be played by a rhythm-and-blues band."

Realizing that simultaneous horn solos would complement his cantilevered piano lines (an "inside," tonally fixated, left-hand bass line opposing an "outside" right-hand line weaving in and out of key, willy nilly), Parker chose soprano saxophone for its African pitch and trumpet because it was the brass instrument closest to soprano in range. To his surprise, he began having problems with American rhythm sections, different in nature but similar in severity to the problems he endured in Paris. His locked-in left hand made a bass superfluous, so he dispensed with it (until forming the Tentet). But drums were too essential to jettison. Unable to find a drummer whose style he considered African enough (though he got a sneak preview of what he was looking for on one job when Michael Carvin forgot to bring his cymbals and played without them), Parker eventually doubled on the drums himself—at least on records, via overdubbing. But because this meant that the Experience sounded better on record than in concert, Parker ultimately abandoned piano in order to give the band the benefit of a real-time, interactive drummer.

The Tentet—formed as a workshop band at the Williamsburg Music Center in Brooklyn in 1983—is an outgrowth of Parker's decision that a music as polyrhythmic as his could do without a piano, but not without drums. "I thought that seven horns instead of two could replace the piano harmonically, with extra power to boot," Parker says. The extra horns also added color and more opportunity for call and response, as well as relieving the bi-tonal monotony that made the Experience trying for even the most patient listener. And because Parker still composed at the keyboard, and frequently assigned the piano's chording duties to guitar, the piano haunted the Tentet as an inaudible presence. A rude-sounding, rhythm-hot big band in the tradition of the Dizzy Gillespie Orchestra with Chano Pozo, the Tentet was also reminiscent of the Algerian horn-and-drum ensembles Parker heard as a child—although he admits this was happenstance: "something I realized only after the fact."

The Tentet's personnel solidified "through a simple process of elimination. Some musicians who were technically equipped to play my music were psychologically unequipped to deal with it. I'll give you an example. On baritone saxophone, I wanted Howard Johnson, who can extend the range of the horn two half-steps at either end, which is an ability that is extremely valuable to an arranger. So I asked him to come to a rehearsal. He did. Everything sounded fine. Then we played my arrangement of 'Lush Life,' and when it was his turn to solo, he stopped dead. I said, 'What's wrong?' He said, '"Lush Life" is a ballad. You can't play it fast.' I said, 'Of course you can. You've been playing it fast for the last five minutes.' He said, 'Yes, but "Lush Life" is like the Bible. You can't change the Bible.' I said, 'But if you know the Bible by heart, you can read it backwards, and it will still make sense.' He got defensive and took his sax and left. The younger musicians—the so-called 'Young Lions'—are the only ones who can play my music."

Indeed, if the Tentet worked and recorded more frequently, it might rival the Jazz Messengers and the David Murray Big Band as an incubator of young talent. Parker has recorded the Tentet twice for his own label: *Errol Parker Tentet* (Sahara 1013) from the band's first rehearsal in 1983; and *Errol Parker Live at Wollman Auditorium* (Sahara 1014), from a rare live gig in 1985. Both of these Tentet albums—which feature powerful improvisatory outbursts and fiery, contiguous, high-speed ensembles on Parker's originals and his Africanized arrangements of "Three Blind Mice," Billy Strayhorn's "Lush Life," and Antonio Carlos Jobim's "Chega de Saudade"—are available from Sahara, 1143 First Avenue, New York, New York, 10021; as are *Graffiti* (Sahara 1011), the most infectious of his Experience albums, and *Tribute to Thelonious Monk* (Sahara 1012), the album that substantiates Parker's claims for himself as a scene-stealing solo pianist. (It's the oddest of all the Monk homages, consisting entirely of standards not particularly associated with Monk. But Parker's liner note offers the most convincing rationale for the program: "It is often a temptation for a musician to tamper with familiar objects. . . . When I am lovingly disfiguring vintage standards like 'Blue Moon' or 'Autumn Leaves,' there is no doubt in my mind that I am paying a tribute to Thelonious Monk, who created an art form and a tradition of making the usual out of the familiar.")

In all, Parker has released a dozen albums on Sahara since 1971, pressing between fifteen hundred and two thousand copies of each without ever having to order a second run. He prefers to sell his albums to stores directly, rather than going through a distributor; this assures him a quick return on his investment, but obviously limits his distribution outside of New York. Still collecting royalties from "Lorre" and insurance money from his car accident (and from a 1982 bicycle accident in New York), he makes ends meet by designing album covers for independent labels and subletting one room of his two-bedroom apartment, which he's kept long enough to reap the advantages of rent control. On one of his living room walls, there is a reminder of his interest in graphics: a lurid collage of Warhol, Avedon, and jazz-festival posters, *Artforum* covers, Little Annie Fannie cartoons, and *Playboy* centerfolds with nipples and vaginas discreetly concealed at the request of his current tenant. "And she and I are not even sleeping together! The only one of my roommates I ever slept with never objected!"

Parker is a hustler. It doesn't look like he's starving, but, as with most musicians, his desire for recognition is as much a matter of ego as survival. "Dizzy Gillespie told me, 'In the 1940s, we had the same problems you have now.' But Dizzy, Charlie Parker, Thelonious Monk, Bud Powell, Kenny Clarke, Ray Brown, Milt Jackson, Miles Davis, and the others all appeared on the scene at the same time, and that made them look like they were part of a movement, which is always reassuring, and also practical because, right or wrong, it allows the press to put a bunch of artists under one umbrella," Parker wrote in a letter he sent me along with articles about himself before our interview. "When it became apparent that I was not an offshoot of the avant-garde (unlike the World Saxophone Quartet, the Art Ensemble of Chicago, or David Murray), it became increasingly difficult for the press to admit that I was 'a force of one' and had created my own school. Jazz originals have always been a rarity. Today, they are so scarce that they have become an oddity, something that looks like it doesn't belong."

Parker's comments are self-serving, but they address a larger injustice. The last mass movement to shake jazz to its foundations was the one Gillespie and Charlie Parker spearheaded forty years ago. By contrast, free jazz was a bunch of strays simultaneously forming their own loyal packs. (Before debating the point with me, try to imagine Ornette Coleman and Cecil Taylor, or Albert Ayler and Sun Ra, playing together the way bop's top dogs did.) Fusion was an aberration, and the virtuoso neoclassicism of the latest class of Jazz Messengers and the samurai-funk leanings of the Brooklyn culture club are most likely just passing trends. The combination of a fringe audience and diversity of styles deemed close enough for jazz effectively rules against another broad-based movement like bop. There aren't going to be any new schools, unless one counts every inventive musician as a school of one. Under the circumstances, it's time to start judging music on its lonely merit instead of its relationship to some wished-for zeitgeist. It's easy to overlook someone like Errol Parker, an anomaly because his theories of bi-tonality and polymetrics have no application outside of his own world. Yet Parker is an original, and that shouldn't be to his disadvantage.

(January, 1988)

POSITIVELY CHARGED

Because I agreed to forgo biographical queries I could look up, my conversation with the saxophonist and composer Henry Threadgill took some unusual turns right from the beginning on the slushy January night I visited him in Brooklyn. "I was reading a Tasmanian tale about Creation recently," Threadgill told me, explaining that he was doing research for the libretto of a new piece, tentatively called *Run Silent, Run Deep, Run Loud, Run High*, for eight voices, eight to ten strings, and four horns. He was limping from what he assumed was an old injury sustained in a jeep accident during his army years. (It turned out to be gout.) "The gods asked the first man and woman on earth to select an object out of a basket, and whatever they chose would determine their future existence. They chose an object they were quite familiar with—a picture, I think it was. On the basis of that, the gods said the man and woman would reproduce themselves and die."

Might the object have been a mirror? I suggested, trying to make sense of the symbolism.

"No, it was a picture. But the gods also decreed that the snake would shed its skin and be immortal. You see, the snake is like the music I'm involved with, continuously shedding layers of skin and taking on different forms. You hear what I'm saying? The music that Charlie Parker made was beautiful, but if you say that it was the ultimate form of music, what you're actually saying is that evolution is over, that jazz is dead.

"I've become very interested in Charlie Parker, but not from the standpoint of playing his music or music based on his, because that would result in a *stylized* piece of music rather than something that takes account of the bigger musical picture. I know, from what I've heard by him and read about him, that Charlie Parker was himself evolving when he died, and would have brought his wider perspective to the music's next evolutional stage. That's what interests me. If the music he played during his lifetime was the yin, what would it have sounded like when he went into the yang?"

From Charlie Parker, the conversation drifted to antimatter, "the flip side of matter," as Threadgill defined it—which is where he left me behind (my brain is that of a simple jazz critic, already overlooked with harmolodics and the Lydian Concept of Tonal Organization). But as I dimly comprehend it, antimatter is composed of particles identical to common particles of matter, but oppositely charged. One could argue that Threadgill's small ensembles—tripartite assemblages of winds, strings, and percussion—bear a similar relationship to symphonic orchestras. Instead of dedicating themselves to interpretation, Threadgill's bands aim to prove that autocratic composition and insouciant collective improvisation can occur simultaneously, in rowdy fraternity, with neither giving an inch. That's the opposite charge.

On one level, Threadgill's sextet (six members plus the leader—remember the fun that Paul Desmond used to have with squares who asked him "How many of you are there in the quartet?") is a vehicle for his writing. In retrospect, so was Air, the much-decorated cooperative trio that Threadgill formed with the bassist Fred Hopkins and the drummer Steve McCall in Chicago in the mid-seventies. (Air remains semi-active, with Pheeroan akLaff in place of McCall.) But in both cases, composition has proved to be a hands-on proposition. "I've been quoted as saying that ninety per cent of Air's music was composed, contrary to what people assumed," says Threadgill, who was born in Chicago on February 15, 1944. His résumé reads pretty much as expected for an AACM alum: an apprenticeship with Muhal Richard Abrams, house gigs at blues clubs and participation in marches and parades, study at Wilson Junior College, and a stint in the military that he says removed civilian hassles and allowed total concentration on music. What's different about Threadgill's background, though, is the period he spent arranging and performing liturgical music for Church of God services, an involvement about which he is elliptical, beyond admitting that it was spiritual as well as professional. "What I meant was that at least ninety per cent was represented on paper in some form, solos and everything, but how it was *executed* was another matter. There was a subtle, almost undetectable line between no improvisation and total improvisation, a unity between the music and the music born of it."

Insofar as it coincided with an end-of-decade *recherche du tempo perdu*, the widespread critical acclaim for *Air Lore* (now available only on compact disc: RCA Bluebird 6578-2-RB), Air's 1979 album of Scott Joplin and Jelly Roll Morton compositions, was based on a misunderstanding. The same impulse that prompted *Air Lore*—the realization that great jazz composers shouldn't have to be physically present in order for their works to be performed—should have also triggered the realization that composition, an insufficiently acknowledged force in jazz at least since Morton's time, didn't have to remain a dirty secret any longer. The plaudits for Air as a unit deflected attention from Threadgill, its most prolific and ambitious composer as well as its most commanding soloist.

After three albums' worth of writing that was technicolor in grandeur and cinemascope in reach by the Henry Threadgill Sextet (1982's *When Was That?*, 1983's *Just the Facts and Pass the Bucket*, and last year's *Subject to Change**), it's no longer possible for even the most myopic of critics to overlook Threadgill the *auteur*. "It's the chicken or the egg question," he laughed, in response to my question about which was more important to him in selecting the sextet's personnel—instrumentation or the instrumentalists he wanted to play with (most jazz composers would answer, the latter). "When I compose for the Sextet, I always have the styles of the individual players in mind. It's a concerto style of writing, even if I don't call my pieces concertos. But in the beginning, I knew what orchestral territory I wanted to cover—and I do mean *orchestral*."

Still, he said, "a continuity and openness between my intentions and the intentions of the individuals in the sextet is the important thing. In some improvised music, there isn't that continuity—a theme is stated, then the solos are almost like an escape from the theme. I'm not making value judgments, but I've never been interested in that sort of approach; not with Air, not with the sextet, and not in the future. I want one fabric, with improvisation as the design that keeps going deeper and deeper into the fabric.

"Critics have written that the three sextet albums on About Time formed a trilogy about death and everything on the other side of death," Threadgill said, without mentioning that he helped to foster that impression with comically morbid cover motifs, titles like "Soft Suicide at the Baths," "Cremation," and "Higher Places," and interview pronouncements to the effect that the further away Americans moved from their graveyards, the further they fled from reality. "That wasn't musical. It was just the subtext, and that's over now. The new album takes up the subtext of antimatter, which was implicit all along."

Antimatter may be matter's flip, but now that Threadgill has rejoined Air's former benefactor, Steve Backer, at RCA Novus, the question is whether RCA will give Threadgill a positive charge. Although ecstatic at the prospect of major-label promotion and distribution, Threadgill must feel like he was through all of this before when Backer signed Air to Arista. What happens to Threadgill when Backer is shown the door, as friends of jazz eventually always are at major labels?

You Know the Number (RCA Novus 3013-1-N) probably won't sell in the box-loads needed to ensure Threadgill's longevity at RCA, but it confirms his preeminence among contemporary composers and bandleaders. It also reveals that *Subject to Change* was aptly titled. An inevitable letdown after the magnificent *Just the Facts and Pass the Bucket*, but one of 1986's best albums

---

* About Time 1004, 1005, and 1007, respectively.

by default, *Subject to Change* was weakened by experiments with text (Pheeroan akLaff's recitation of an excerpt from Emilio Cruz's *Homeostasis*, and guest vocalist Amina Claudine Myers's interpretation of lyrics by singer Cassandra Wilson, Threadgill's former wife) and the less-than-successful integration of the trumpeter Rasul Siddik and the trombonist Ray Anderson as replacements for Olu Dara and Craig Harris, the scene-stealing brass duo on the sextet's first two albums. But *You Know the Number* makes it obvious that the real problem was that the nominal rhythm section was on its way to, but had not quite arrived at, parity with the horns—and that the problem has been resolved. The drummers akLaff and Reggie Nicholson, the bassist Hopkins, and the cellist Deidre Murray are up front for most of "To Be Announced," making a noise so contagious that the layered improvisations by Threadgill, Siddik, and the new trombonist Frank Lacy almost qualify as a bonus.

"To Be Announced" is a delicious near-calypso, but it's probably the album's least interesting cut, which tells you how scintillating the rest is. And the best part of it is that, with all the windows Threadgill leaves for improvisation, this is music resolutely in flux: if it sounds this good on record, it's thrilling to imagine what it'll sound like in concert. "Bermuda Blues," the opener, is one of those riotous affairs that prescribe the blues as the only cure for the blues, with Hopkins evoking the galumphing ghost of Mingus, Threadgill testifying with preacherly fervor, and Siddik guffawing like Dara used to, without forsaking his own Lee Morgan lineage. "Good Times" is as immediately accessible, a jump-up full of surprising extras—Murray's cello counterpointing the homey theme behind Lacy's tailgating trombone solo, for example—and both "Theme from Thomas Cole" and "For Those Who Eat Cookies" are fanfares so flagrantly theatrical they beg for a runway (and perhaps hint at what the opera that Threadgill is writing with Thulani Davis will sound like). But even with so much to choose from, the piece that strikes the most satisfying bargain between composition and improvisation is "Silver and Gold, Baby, Silver and Gold," a gritty, Ayler-like dirge for commiserating alto, bass, and cello, with pins-and-needles percussion and poker-faced brass. If this is antimatter, give me more.

(*February, 1987*)

BLOWING IN
FROM CHICAGO

"In my profession, nothing is more alienating than going to a smash hit and not being able to share the good time everyone seems to be having," Robert Brustein writes in *Who Needs Theatre* (Atlantic Monthly Press), a recent collection of his essays and reviews. I suspect that the feeling he describes is something all critics experience from time to time, and not just those forced to endure bad popular theater. I know it describes how I feel when I'm obliged to attend a sold-out fusion concert: I look around and I'm the only one the beat's not reaching. But jazz critics must be unique in feeling even more isolated and demoralized at performances that meet their highest standards—a quick head count is usually enough to remind us how few people share our appreciation for undiluted jazz. Not long ago, I flew to Chicago to hear Edward Wilkerson, a local thirty-five-year-old saxophonist and composer who had released two impressive albums on his own label but had not yet performed on the East Coast with either of his bands—his twenty-plus-member orchestra, Shadow Vignettes, or his smaller group, 8 Bold Souls. I heard 8 Bold Souls at a loft a few blocks west of the Sears Tower and the river. The concert was as electrifying as I expected it to be, but counting myself and the two small children of one of Wilkerson's sidemen, there were exactly two dozen people there to enjoy it.

The audience for jazz is microscopic all over the country, and local performers are taken for granted everywhere, so the low turnout shouldn't have surprised me. Maybe it did because I was still disoriented from my flight (flying from one big city to another in less than two hours, you don't really *see* anything to indicate that you're in a different part of the country—just airport terminals, expressways, and skyscrapers). Or maybe it was because I was feeling sentimental on my first visit to Chicago, a city as important as New Orleans in the evolution of jazz from the blues, and as important as New York in determining the direction of the contemporary jazz avant-garde. Notwithstanding the post–World War II dixieland revival, the records that such New Orleans émigrés as King Oliver, Louis Armstrong, Jelly Roll Morton, Johnny Dodds, and Jimmie Noone made in Chicago in the twenties provide the best clue we have to what New Orleans jazz sounded like at the beginning of the century. But if Chicago postmark on so much "New Orleans" jazz underscores the folly of geographical designations, so does the New York postmark on much of what has been accepted as pure "Chicago" style: the all-stops-out records that Eddie Condon, Bud Freeman, Jess Stacy, and other white midwesterners made in the thirties and forties, after relocating East. New York's stranglehold on all aspects of the music business by the end of the twenties ensured that Chicago's reign as the jazz capitol would be brief (so, on a strictly symbolic level, did Armstrong's move to New York in 1929). In the decades since, Chicago has produced its share of pacesetters, including the boogie-woogie pianists Albert Ammons, Jimmy Yancey, and Meade Lux Lewis in the thirties, and Sun Ra in the fifties. But these Chicago musicians achieved widespread recognition only after establishing themselves in New York, and this is still the rule, despite an unexpected development of the late 1960s.

That was when the names Roscoe Mitchell, Lester Bowie, Joseph Jarman, Muhal Richard Abrams, and Anthony Braxton first began to appear in *down beat*, usually accompanied by the letters AACM (the Association for the Advancement of Creative Musicians). According to *down beat*'s Chicago contributors, the AACM represented the next logical step after Ornette Coleman and John Coltrane. What supposedly made the AACM different from the "free" jazz being played elsewhere was its greater emphasis on dynamics, texture, sound color, and the soloist's role within the ensemble. At first, many of us in the East, suspecting the Chicago-based *down beat* of local boosterism, read about these AACM members with amused skepticism. If they were that good, the logic went, why were they still in Chicago? And why hadn't they been recorded? They soon were, but it wasn't until members of the AACM and its sister organization, BAG (the Black Artists Group, of St. Louis) hit the road—to Europe at the end of the sixties, and, more to the point, to New York in the next decade—that the significance of events in the Midwest in the late sixties was generally acknowledged.

About five years ago, another unfamiliar name began to turn up in the dispatches of the same Chicago critics who had been early advocates of the AACM. In his 1984 book *The Freedom Principle* (Morrow), John Litweiler described Edward Wilkerson, Jr., a younger AACM member, as a "remarkable saxophonist [who] takes the funky jump band tenor idiom into Free territory with infectious rhythms and a full, huge sound." At that point, Wilkerson had recorded three import albums with the Ethnic Heritage Ensemble, a loose-as-a-goose trio led by the percussionist Kahil El'Zabar. These offered little to substantiate Litweiler's praise, aside from a chuckling solo on Wilkerson's own "A Serious Pun"—aptly titled, given the mock gravity of its étude-like unison saxophone theme. (It's included on *Three Gentlemen from Chikago*, Moers Music 01076.) And when, in 1986, Wilkerson released his own album, *Birth of a Notion* (Sessoms Records 0001), by his Shadow Vignettes, it offered no way to judge his prowess as a saxophonist, for he laid down his horns in favor of a composer's scoresheet and a conductor's baton. But the album—a do-it-yourself project produced for under six thousand dollars when Wilkerson realized that it made as much sense to put out his own record as long as he was going to the expense of making a demo—marked him as a composer to be reckoned with.

The writing on *Birth of a Notion* was jocular, robust, eclectic, and deceptively improvisatory in tone, and what pulled you into it right away was Wilkerson's redemption of cliché, best demonstrated on "Honky Tonk Bud" and "Quiet Resolution," the two tracks with vocals. Although independently produced albums by performers without New York credentials tend to be overlooked, *Birth of a Notion* started a buzz about Wilkerson outside Chicago, but not as unanimous a buzz as he deserved. Too many influential critics ignored the album, and the reaction among those who reviewed it was not wholly favorable, the vocal tracks mentioned above being the biggest stumbling blocks. The lyrics to "Quiet Resolution" are hopelessly banal—"Our love is different from all the others / 'Cause our love is holding up the planet earth / And keeping if from self-destructing / I love you so, you know?"—and a British critic denounced Rita Warford's poker-faced vocal as "some of the worst jazz singing" he had ever heard. The impossibly rich string writing behind Warford (I think of it as "Radio City Music Hall," and Wilkerson has described it as "Disneyland") should have made it clear that Warford and Wilkerson were having fun at the expense of early-seventies cosmic jazz, but the joke was lost on anyone lacking a sense of humor. The actor John Toles-Bey, who has since appeared in *Weeds* and *Midnight Run*, narrates "Honky Tonk Bud," the story of a small-time South Side drug dealer who becomes a folk hero when he's nabbed in a FBI sting. As delivered by Toles-Bey, "Honky Tonk Bud" is flamboyant theater; and behind him, Wilkerson wheels in orchestra motifs as though they were stage props, establishing the ghetto milieu with a bass line borrowed from the old Cannonball Adderley hit, "Sack of Woe," conveying Bud's inflated self-importance with helium brass riffs, and underscoring his arrest and trial with droll send-ups of TV crime-show themes. In other words, there's enough going on in "Honky Tonk Bud" to keep you riveted even if the story it tells strikes you as representative of an infatuation with street culture that jazz should have outgrown in the era of *Shaft* and *Superfly*. But some reviewers dismissed "Bud" as a middlebrow attempt at rap, even though the giveaway that Wilkerson wasn't trying to top teenage rappers at their own game, as a commercially oriented jazz musician might, was the absence of a beat-box rhythm track. Besides, "Honky Tonk Bud" wasn't even a rap. It was a "jailhouse toast"—a black inmates' rhymed equivalent of an urban legend—but few jazz critics knew enough about contemporary black vocal styles to perceive the difference. (And those who found the narrative not "streetwise" enough might be surprised to learn that David "Swif-a-rony" Smith, who is credited with Toles-Bey for adapting the lyric, was gunned down in a drug deal shortly after the album's release.)*

Birth of a Notion made me eager to hear 8 Bold Souls, which my Chicago colleagues said was an even better showcase for Wilkerson than Shadow Vignettes. The release of *8 Bold Souls* at the tail end of 1987 proved them right (Sessoms 0002, available, as is *Birth of a Notion* from Sessoms Records, P.O. Box 6812, Chicago, Illinois 60680). Not only does Wilkerson actively participate as an instrumentalist, but the group also gives more conclusive testimony to his resourcefulness as a composer able to draw comparable orchestral richness out of a smaller number of horns. In addition to playing clarinet (and the seldom-heard alto clarinet) on the ensemble passages, Wilkerson solos on both alto and tenor saxophone, offering the delightful paradox of short, brusque, exclamatory phrases delivered in what can only be described as a croon. His tenor solos are especially notable for their conjunction of imposing muscularity and cool deliberation: the sound of rhythm 'n' blues gone abstract, which one also hears in the solos of Chicago tenor saxophonists as diverse as Gene Ammons, Von Freeman, John Gilmore, Johnny Griffin, Eddie Harris, and Joseph Jarman.

With the bassist Richard "Jess" Brown, the cellist Naomi Millender, and the tuba player Aaron Dodd in 8 Bold Souls' rhythm section, Wilkerson can get as many as three bass lines going at once, and his music becomes deliriously heavy-bottomed when the versatile Mwata Bowden plays baritone saxophone. But part of the fun of listening to 8 Bold Souls is in guessing where the tuba and cello will show up next. These instruments give Wilkerson two more "horns" to

---

* Also recommended: Scott Laster's *Honky Tonk Bud*, a ten-minute stereo video that induces nostalgia for the days before Chuck Norris and Avenging Angels, when black exploitation movies starred black actors. It's available from Urban Legend Film, 646 West Webster, Chicago, Illinois 60614. It's also included—along with Peter Bodge's animated *Bird Lives*, D. A. Pennebaker's *Daybreak Express* (with music by Duke Ellington) and Robert Mugge's *Is That Jazz?* (with Gil Scott-Heron)—on *Jazz Shorts*, available from Rhapsody Films, P.O. Box 179, New York, N.Y. 10014. Hearing "Honky Tonk Bud" and seeing the video, one isn't surprised to learn that Wilkerson, an avid collector of vintage cartoons and big band "soundies," is ambitious to score a film. "I can't say I've already done a score," he says, "because Scott shot *Honky Tonk Bud* to my music, not the other way around."

work with, and he frequently has one or both playing lead. There are six Wilkerson compositions on the album, each with something fresh and inventive, although the collective improvisation on the dirgelike "Through the Drapes" is a mite too careful to deliver its intended impact. "Chapel Hill" is a beautifully sustained tone poem, with more than a hint of a march in its suspenseful ostinato rhythm. It's also the track that, in blurring composition and improvisation, most suggests Henry Threadgill, a former Chicagoan, who seems to be Wilkerson's primary influence. "Favorite Son" successfully translates a funk afterbeat into a broken $\frac{7}{4}$ meter, thanks to the aplomb with which the drummer Dashun Mosley underlines the alternating moments of buoyancy and constraint in Wilkerson's alto solo. With their mulling preludes and explosive main themes, "The Hunt" and "Dervish" are enjoyable for their sheer variety. The latter, to give you some idea, overlaps what could pass for an Ennio Morricone spaghetti-western theme, a fast bebop lick with bass and cello counterpoint, a stop-time Wilkerson tenor solo violent with split tones and overblown notes, a blues riff that recalls Thelonious Monk's "Straight, No Chaser," and brass codas by the trombonist Isaiah S. Jackson and the trumpeter Robert Griffin, Jr., that maintain the almost unbearable level of intensity reached in Wilkerson's solo. The album's boldest stroke is "Shining Waters," with Mosley's safari rhythms, Dodd's baying tuba, and a growling Bowden clarinet solo pushing the Duke Ellington "jungle" orchestra of the 1930s deeper into the bush. Like "The Names Have Been Changed" from *Birth of a Notion*, "Shining Waters" resembles "Bloodlines," "Mama and Papa" and other back-to-the-future big-band flag-wavers by Wilkerson's former teacher Muhal Richard Abrams, in attempting to recapture the "spark" (as Wilkerson has put it) of the big bands of the thirties and forties without lapsing into period re-creation. But Wilkerson has realized this goal to an even greater extent than his mentor; his pieces in this vein dial up both the erudition and the showmanship of such seminal black orchestrators as Ellington, Don Redman, and Fletcher Henderson.

I heard 8 Bold Souls at Modalisque, a vintage-clothing store by day that—with its window shades, exposed pipes, and clutter of forties and fifties kitsch (including sofas and love seats to stretch out on when the folding chairs got uncomfortable)—provided the sort of agreeably funky concert setting that hardly exists in New York anymore. This was the third in a series of four consecutive Sunday night performances by 8 Bold Souls; the sort of arrangement that Wilkerson prefers, because it gives him a chance to put his music through a workshop process. The first concert in the series, two weeks earlier, drew an overflow crowd from the Chicago Jazz Festival, which had ended just hours before in Grant Park. People were still talking about how the police raided Modalisque that night, confiscating all the money taken in at the door—the owner hadn't realized that she needed to apply to City Hall for a license to put on live entertainment, and another in order to sell beer. The crowd wound up passing the hat for Wilkerson and his musicians. The small audience the night I heard 8 Bold Souls was more typical of their drawing power, I was told. Also playing in Chicago that night were the Leaders, an all-star sextet including the trumpeter Lester Bowie, the tenor saxophonist Chico Freeman, and the percussionist Famoudou Don Moye—three AACM members with national reputations—and this competition for the jazz dollar may have been a factor in the poor turnout for 8 Bold Souls (the Leaders were all anyone was talking about earlier that day in the Jazz Record Mart, and Wilkerson himself regretted missing their shows).

Beyond the lack of hometown support for Wilkerson, I found nothing to complain about at the concert. In *Who Needs Theatre*, Brustein calls theater, the symphony, opera, and dance "our last chance to see animate human beings engaged in a collective creative act." What about jazz and pop? In addition to pieces from *8 Bold Souls*, the group played a few of Wilkerson's more recent compositions, and even if these frequently seemed tentative in execution, there was satisfaction for the listener in hearing eight musicians feeling one another out on works in progress. In concert, the pieces from the record were given more extended treatments, in which Wilkerson allowed the soloists (himself included) greater freedom. But in the improvised solos, there was none of the gratuitous muscle-flexing that vitiates too much contemporary jazz. The relationship between Wilkerson's themes and even the lengthiest solos was clear. The small audience showed genuine sophistication in reserving its applause until the end of each piece, instead of clapping for individual solos as jazz audiences customarily do to mark time while waiting for the next soloist to "get hot." The charter members of the AACM avoided the excesses of the New York "energy" players but substituted excesses of their own, in the form of long, unaccompanied solos and sometimes nearly inaudible ensemble percussion passages. In Wilkerson's band, there's no wasted motion at all.

The only missing element was the theatricality that I read was an integral part of Wilkerson's live presentation. It's supposed to be what links him to a Chicago tradition epitomized by the AACM's face paint and Sun Ra's flamboyant robes and parades through the auditorium. Over lunch the next day, Wilkerson assured me that Shadow Vignettes was the more theatrical of his two bands. "I like to use the whole auditorium, not just the stage," he said, explaining that during live performances of "Quiet Resolution," for instance, he often has the soprano saxophonist Light Henry Huff rise from the audience and pull his horn from beneath his overcoat at precisely the moment that Rita Warford wonders aloud, "Where's my cosmic lover?"

Somewhat shy on stage himself, the compact and powerful-looking Wilkerson nonetheless enjoys having a three-ring circus going on around him. He was born in Terre Haute, Indiana, and raised in Cleveland and its affluent liberal suburb, Shaker Heights (the well-regarded and voluntarily integrated system attracted his parents). He is the only musician in his family: his father is a physician, his mother is a housewife, and his three sisters are lawyers or MBAs. Wilkerson began his association with the AACM in 1971, when he enrolled at the University of Chicago. Although he earned his bachelor's in music theory and analysis at the university, he considers his studies with Abrams to have been of more practical value, because of Abrams' "hands-on" approach to music ("He would discuss a particular concept, then have his students write a short piece utilizing it right away") and his focus on "the principles underlining all music regardless of idiom, period, or style. That was important, because, in school, we never examined twentieth-century music, much less music by black composers. When I left Cleveland, I knew I wanted a career in music, but I didn't know where to start. There are a lot of fables about how you're supposed to practice, sit in all the time, other musicians will hire you, and before you know it, you're leading your own band. But I knew that was true for maybe one person in a hundred. The AACM was an alternative route. Muhal held classes every Saturday in AACM headquarters, on the second floor of a nursery school. Classes were three hours long. You paid a dollar, which went toward refreshments. The people there already knew their instruments. Most of the students were my age or older: George Lewis, Chico Freeman, Douglas Ewert, Amina Claudine Myers. Most had just finished college."

Wilkerson himself now teaches at the AACM's school on the South Side, and serves on the organization's governing board. The size of his bands and the challenging nature of his music preclude the regular work in neighborhood lounges that has traditionally sustained stay-at-home musicians. He is single, so it is probably his commitment to the AACM and the benefits he derives in return that explain why he hasn't already abandoned Chicago for New York. "You can always call on fellow AACM members to perform your music, and I appreciate that, because the greatest problem for any contemporary composer, regardless of idiom, is hearing his music performed. Take the Chicago Symphony. You can count the number of times a season they play a living American composer on the fingers of one hand. I don't know what I'd do if I were composing in that idiom."

Though he now mostly works with his own bands and El'Zabar's Ethnic

Heritage Ensemble, he's paid his dues backing blues singers (including Albert King and Bobby "Blue" Bland) and oldies acts (including Gene Chandler and Little Anthony and the Imperials). "But those weren't just bread gigs," he says, "because, as a matter of fact, the money wasn't even that good. Playing behind Gene Chandler was a thrill. It was intense. I mean, he was The Duke of Earl! And Little Anthony's songs, I got goose bumps just hearing them again. He's a very sophisticated musician, a big Coltrane fan. Most of those R & B guys from the fifties and sixties know many types of music, not just the music they're associated with, and you can learn a lot about showmanship from them. They remember when there was a lot more melding of music than there is now, when you might have a vocal group, a jazz band, a comedian, and dancers, all on the same bill."

I first met Wilkerson when he visited the East Coast to canvass critics and radio stations, shortly after releasing *Birth of a Notion*. Although this promotional tour was possible only because Wilkerson took advantage of supersaver flights and the generosity of family and friends in New York, Philadelphia, Boston, and Washington, he told me that he wanted people to think he was blowing his life savings. He hoped to drum up sufficient interest in Shadow Vignettes to bring the band East for an extended tour. "Once you rent a Greyhound, it doesn't matter if you have sixteen pieces or forty," he told me. "The cost is the same. I have a friend in Chicago, a baritone saxophonist, who went on the road with Cootie Williams in the 1940s, for a tour that was supposed to last two weeks. But the band was actually on the road for a year and a half. They never made enough money to go home, so they had to take jobs all over the South and Midwest, and they spent all the money they made. My friend remembers it as the best period of his life. He said that the friendships you form in a group that large that's together that long are something special, and that it's a shame modern players are denied that kind of experience." When I asked him why he didn't play with Shadow Vignettes, he explained that writing for and conducting such an oversized ensemble was all he could handle at the present time. "Maybe eventually. But right now, I look at myself as the intermediary between the band and the audience. That's how many of the big band leaders I see in the soundies acted. Willie Bryant, for example. He would do things to ignite the band, but almost in the way that somebody in the audience would. Duke, too, although he played. He always acted as a kind of host, and that's the role I play on stage with Shadow Vignettes."

The tour never happened, but Wilkerson remained optimistic when I caught up with him in Chicago. "There's a change happening in music," he told me. "You're going to have bands in cities all over the country, with their own distinct regional sounds, achieving national and even international renown, just like the territory bands that every region of the country had in the twenties and thirties."

He has a point. What used to lure musicians (including many from those territory bands) to New York was the promise of steady employment and even stardom represented by that city's nightclubs and independent jazz labels. But New York's nightclubs no longer provide enough work to go around, and most of those labels are no longer active. In 1989, a musician in another city isn't necessarily any worse off for staying put. Jazz is becoming decentralized, a process that was already in motion twenty years ago, when the AACM asserted its independence from New York. Along with Wilkerson, the tenor saxophonist and composer Paul Cram (Toronto) the trumpeters Dennis Gonzalez (Dallas) and Paul Smoker (Cedar Rapids), and Boston's Either/Orchestra are among those who deserve mention for forging original styles in far-flung places. There's just one problem: a handful of New York–based critics still shape the opinions of others around the country. If there's anything more depressing than the microscopic size of the jazz audience, it's the geocentricity of these Manhattan tastemakers. Although Wilkerson has performed with his bands throughout the Midwest and taken them to European festivals, his lack of exposure in New York has prevented him from receiving his full share of acclaim. Happily, this might soon be changing: 8 Bold Souls finally made its New York debut last November, and Shadow Vignettes is scheduled to perform at the Brooklyn Academy of Music later this year, as part of New Music America 1989. I suspect that we're going to be hearing a lot more about Edward Wilkerson. His combined skills as a bandleader, composer, and soloist put him on an equal footing with Anthony Davis, David Murray, Henry Threadgill— the best and most visible of his contemporaries in New York.

(February, 1989)

COMPOSITION
AFTER THE FACT

The music of the spaghetti-western soundtrack composer Ennio Morricone (Bernard Herrmann to Sergio Leone's Hitchcock) as interpreted by the Lower East Side improvisational gamemaster John Zorn (*Archery, Pool, Lacrosse*) hardly sounds like an appetizing prospect. It isn't that Morricone is unworthy of homage, just that an arch-experimentalist like Zorn would seem incapable of doing justice (except as sour parody) to such a juicy sensualist. I confess that the bond that certain of my colleagues feel with taking-it-out-of-context eclectics like Zorn, Kip Hanrahan, Bill Laswell, and Hal Wilner puzzles me, unless the answer has something to do with their mistaking the fringes for the cutting edge—the desire of both performers and critics to be in the vanguard, even if the vanguard of *what* is impossible to say. Critics tend to gravitate to performers who've reached the same conclusions they have, and the conclusion, in this case, would seem to be that because bebop is bankrupt, rock 'n' roll is moribund, and the classical-music avant-garde ain't what it used to be, recombinant forms are the only honorable alternative (and hip hop, Brazilian, tango, and shakuhachi are the current flavors of the month).

But, surprise!—few recent albums deliver as many epiphanies, or as much sheer fun, as *The Big Gundown* (Icon/Nonesuch 9 79139-1F), Zorn's deconstruction of Morricone. Or maybe it's not such a surprise after all, given the talent for working from palimpsests that Zorn displayed on *That's the Way I Feel Now* and *Lost in the Stars*, Wilner's salutes to Thelonious Monk and Kurt Weill, respectively. To be sure, Zorn takes liberties with the material. But as film music, the material was crafted to withstand endless variation (as was also true in the case of *Amarcord Nino Rota*, Wilner's most successful project), and Zorn generally resists the artsy ploy of going for the Big Frisson, of merely striking coy or anguished poses in the general vicinity of Morricone. His interpretations, faithful in their fashion, pass muster as full-blown recompositions, with a linear vitality often lacking in his own scratch improvisations for ensemble, and with a higher music-to-noise ratio than he usually favors.

The Big Gundown registers as Zorn's most personal work, even though only one of its ten tracks is his ("Tre Nel 5000," with a tremulous opening that sounds more like film music—that is, pastiche—than anything else on the album), and even though he's present as an instrumentalist (alto saxophone and game calls) on only that track and five others. Perhaps because he shares Morricone's love for guitar whine and found sound, *The Big Gundown*'s most euphoric cuts are those with generous amounts of Third-Stone-from-the-Sun rave ups and things going bump in the night. The guitarists Jody Harris and Robert Quine duel like aging gunfighters on the evocative but somewhat overlong "Once upon a Time in the West," and Harris, Fred Frith, and Arto Lindsay bounce off one another irresistibly on the jaunty "Milano Odea," which at a compact 2:57 would be a hit single in a fairer world than this one (the flip side would be "Poverty" from *Once upon a Time in America*, with Toots Thielemans's doleful yet sweetly nostalgic whistling).

Guitarists also figure prominently on tracks where other instruments supply the impetus. On "Erotico" (from *The Burglers*), Bill Frisell and the drummer Bobby Previte team with the organist Big John Patton to postmodernize the Blue Note sound behind Laura Biscotto's "sexy Italian vocal" (try to imagine Yma Sumac in heat). And along with Mark Miller's lashing drumbeats, Wayne Hurvitz's pounding piano ostinato, and Diamanda ("Wild Women with Steak Knives") Galas's demented, oppressively erotic cries, Vernon Reid's brutal downstrokes contribute to the overwhelming air of Catholic self-flagellation on "Metamorfosi" (from Morricone's score for *The Working Class Goes to Heaven*, though, from the sound of things, their purpose for being there is to toil in sweatshops owned by the idle rich, just as on earth).

The participation of straight-lifes like Theilemans and Patton in this project is surprising, given that the rest of the personnel reads like a Bohemian Art Music Who's Who. But one of *The Big Gundown*'s consistent joys is the redemptive use that Zorn makes of such déclassé instruments as organ, harmonica, and accordion. Collage is one of his strategies, employed to startling effect on "Battle of Algiers," where Anton Fier's military drumming underscores the collision between Vicki Bodner's sylvan English horn and Christian Marclay's urban turntable screech; on the crafty "Giù la Testa" (from *Duck You Sucker!*), which begins with sitar-like synthesizer drone and over-blown shakuhachi and samisen and ends with goofy Eurosong "shoop shoops"; on "Peur sur la Ville," with pounding heartbeats, Tim Berne's alto saxophone squalls, and the sound of breaking glass; and on the title track, with its *batucada* ensemble, *Rawhide* and *Für Elise* interpolations, claustrophobic wordless vocals, and galloping, wide-open-spaces percussion.

"The studio . . . is a way of composing, of documenting music on tape rather than score paper," Zorn is quoted in the liner notes. "[Even] with all the graphic notation of the sixties, and all the other alternative notations, including my own game scores, the most revolutionary idea is being able to get these musicians, whose sounds cannot be written down, notated on tape." In other words, Zorn discovered, in the process of recording *The Big Gundown*, that composition could come after the fact—old news for pop record producers, perhaps, but a breakthrough for American avant-gardists like Zorn, who (unless affiliated with universities like Princeton and Dartmouth) have been cate-gorically denied access to state-of-the-art recording technology. It's instructive to compare Zorn's interpretations with Morricone's originals, but even more instructive to compare Zorn's various drafts. Toward the end of 1984, Yale Evelev, Zorn's producer, sent me an unmastered cassette of some of the tracks on *The Big Gundown*, including "Peur sur la Ville" and "Giù la Testa." The raw tapes sounded like studio jams with Morricone as a tenuous reference point, but in the two years since, Zorn has worked wonders with this music, layering more instruments on top of those he already had, enlarging some parts and diminishing others, generally bringing everything into sharper focus.* He deserves another trip to the studio, for his own music this time. His ministrations on behalf of Morricone auger well for his future ministrations on behalf of Zorn.**

In comparing *The Big Gundown* to Zorn's own scores, I might be at a disadvantage from knowing those scores only on record—a point driven home by my mixed feelings about Butch Morris's new album, *Current Trends in Racism in Modern America (A Work in Progress)* (Sound Aspects 4010). Morris, a gifted enough trumpeter on the evidence of his early work with Frank Lowe and the David Murray Octet, is among a growing number of instrumentalists who are laying down their horns in order to conduct and compose for large ensembles—though in Morris's case, the words "conduct" and "compose" are somewhat misleading, because what he actually does, with Murray's Big Band as well as with his own groups, is to "cue" improvisation. Morris is thus something of an anomaly in jazz, where leadership has traditionally involved instrumental laying on of hands; and even more so in his chosen field of free improvisation, where a conductor would seem to be as out of place as an omniscient narrator in the novels of William Gass or John Barth. Yet a long, slowly evolving piece I heard Morris conduct in Philadelphia last summer vindicated his approach; the authorship of the music was never in question,

---

* On the other hand, it's possible that I'm reading too much into a better mix. Zorn has written: "People often assume that the fast changes in my music are a result of tape editing. Nothing could be further from the truth. . . . There are no tape edits in my music, ever. We rehearse section one, then record it, rehearse section two, then record it. . . ." This would seem to imply that he doesn't overdub, either.

** Also recommended: Zorn's 1987 album, *Spillane* (Nonesuch 9 79172-1). On first impression, the title piece sounds like an edgy flip through the world's hippest and most diversified record collection, with snippets of hick-a-billy, Chicago blues, cool jazz, trombone-heavy salsa, and Nelson Riddle television themes (all this plus blood-chilling screams and Lounge Lizard John Lurie's tough-guys-don't-enunciate narration of an unsavory text by Arto Lindsay: "There was a party of bullets going on inside her shirt. . . . There are only so many ways a woman can undress. I thought I had seen them all"). But on repeated listening, the fragments fall together beautifully, and the work gradually acquires the staying power of a bad dream. The Nonesuch album also includes Zorn's flamboyant concerto for the bluesman Albert Collins and an ambitious but rather shapeless homage to the late Japanese film star Ishihara Yujiro, featuring the Kronos String Quartet, a turntable scratcher, and a cooing Japanese singer. Because I find the alto tone that Zorn affects when playing "straight" jazz to be unnecessarily harsh, I am less enthusiastic about *The Sonny Clark Memorial Quartet* (Black Saint BSR-0109).

even though Morris neglected to sign it (in effect) by sitting down at the piano or bringing his trumpet up to his lips.

By comparison, *Current Trends in Racism In Modern America*—an album-length work in two parts recorded in concert at the Kitchen in New York two years ago by a ten-member ensemble including Lowe, Zorn, and Marclay—is static and inchoate, lacking the euphonic curve I remember from Philadelphia. This may be because the record documents Morris's first "conduction" (if I read his Professor Irwin Corey–like liner notes correctly) and it took some time for the parts to fall into place, or because Morris felt that euphony was inappropriate for a piece about racism, or merely because I miss the collegial spirit that comes from hearing music as part of a live audience. More likely, it's because the compressed grooves of a phonograph record are a poor medium for free improvisation, with its dynamic swells and decays and absence of familiar reference points (does this explain the fact that the leading free improvisers have been British Marxists with an instinctive distrust of mass reproduction?). For all that, *Current Trends* is as satisfying a free improvisation (or shall we say "semi-free," in light of Morris's calling all the shots and starting off, I would suppose, with some outline in his head) as one is likely to hear on record, with an abundance of passages almost dreamlike in their celerity, and some inspired Zorn/Lowe saxophone keening in part two (which sounds more like an encore than a continuation). Still, in view of the wonders that Zorn performs in the studio on *The Big Gundown*, one suspects that Morris might be capable of a similar transformation if afforded the same largesse. Undoubtedly, *The Big Gundown* and *Current Trends in Racism in Modern America* were made possible in part by a fistful of dollars from the usual public funding agencies—it quickens the pulse to imagine Zorn and Morris (and others like them) could come up with for a few dollars more.

(January, 1987)

instrumentalists

SUNDAY BEST

Sweet Basil can be as uncomfortable as any other Village nightclub when overcrowded, which it almost always is. Still, it's the only one I'd ever want to be stuck in during the day, maybe because it's above ground with a glass enclosure creating at least the illusion of fresh air, a restaurant rather than a bar, with wood paneling and overhanging plants and an entirely more agreeable odor than, say, the Village Vanguard's (time to change the kitty litter, Max). I enjoy Sweet Basil most on Sunday afternoons, when the price of an omelette also buys the opportunity to hear Doc Cheatham.

With most of his contemporaries dead or impaired, the eighty-one-year-old Cheatham pulls a large sentimental vote as the last of a vanishing breed—the hot, lyrical, swing-era trumpeter. But that only partly explains his allure. His tonal abundance would be praiseworthy for a much younger player; so would his charred low notes, his comedic plunger and buzzing mute, his melodic dash, and the resonant shake he can give a blue note. To say that an elderly musician is playing better than ever usually betrays wishful thinking, if not egregious condescension. But in Cheatham's case, there's little evidence to the contrary. In his prime, with the Cab Calloway Orchestra in the thirties, fellow musicians recognized him as an ace lead. "The lead trumpet swings the band," he remarked testily, lighting a cigar, between sets at Sweet Basil one Sunday. I'd made the mistake of saying that everyone knows the drummer swings the band but hardly anyone knows what the lead trumpeter's duties entail. "If the lead trumpet isn't doing his job, it doesn't matter what the drummer does. Even Chick Webb would sit elevated behind the brass section so he could look at the lead trumpet chart and copy the licks on his drums. That's where people got the idea that it's the drummer who swings the band." But because Cheatham was so good a lead, he was typecast as an indispensable section man and rarely tapped to solo. "I didn't know any better," he told me. "I thought that being lead trumpet in a big band was the greatest thing in the world, and I thought the big bands would be around forever."

Cheatham wears spectacles and combs his stringy hair forward over his brow, Julius Caesar–style (it might be a toupee). His reddish complexion and high cheekbones suggest American Indian heritage. He was born and raised in Nashville, a member of a family he describes as "ordinary," although it's clear they were black bourgeoisie by turn-of-the-century standards. His father was a barber "who catered to an all-white trade in his own shop in the middle of town. That bothered some people who tried to force him out. But a friend of his, an Italian immigrant who came over here with nothing but made a fortune in the candy business with my father's encouragement, bought the building so he could stay." His mother was a schoolteacher, and most other family members were medical professionals, which explains how Adolphus Anthony Cheatham was nicknamed "Doc"—it was a futile nudge from his parents, who were none too happy with his decision to become a musician.

Cheatham's early idols were Charlie Craft and George Jackson, now-forgotten riverboat trumpeters. "But I never tried to blow like them, because I couldn't. I leaned toward the sweeter side of trumpet playing"—exemplified by Arthur Whetsol, Duke Ellington's lead trumpeter, who became Cheatham's role model. After knocking around with territory bands for a decade or so, and journeying to Europe with Sam Woodling ("the first job that really counted for anything"), Cheatham joined Calloway in 1932 and quit in 1939 "because I was just run down. Seven years, that's a long time to be jumping in and out of buses, not eating right, not sleeping right."

After a period of illness, he played in the short-lived big bands led by Benny Carter and Teddy Wilson, neither of which afforded him substantially more solo opportunity than he had gotten with Calloway. Even as a member of Eddie Heywood's sleek Café Society sextet, he was eclipsed by the more mercurial Lem Davis and Vic Dickenson. In the fifties and sixties, when jazz gigs were scarce, Cheatham free-lanced with society orchestras and Latin bands. By the mid-seventies, he was ready to pack his horn away for good, when luck finally intervened. It's difficult to say what did the trick, but about ten years ago, after several concerts with the New York Jazz Repertory Company and appearances at numerous European festivals, everyone was raving about Doc Cheatham as though he was a new discovery—which indeed he was.

Cheatham's Sunday brunches at Sweet Basil, now in their seventh year, draw a wider generational spread that the mostly thirtyish crowd that frequents the club for modern jazz at night, and natives seem to outnumber tourists for a change (there is also a Saturday brunch, with the tenor saxophonist Eddie

Chamblee). This is New York leisure as those of us who read George Frazier at an impressionable age persist in imagining it: folks long past having to fight for their right to party, brazen enough to brunch from three to seven in the afternoon, after staying up all night and lying in bed till noon with the Sunday *Times*. Fantasy, I know. But the atmosphere is so relaxed that nobody much notices when one besotted fool who seems to be there every week sticks out his chest and cakewalks down the aisles to Cheatham's insinuating medium tempos.

Cheatham seizes the opportunity to enjoy himself, too, circulating the room and greeting regulars between sets. On stage he's irrepressible, waving to friends or using his hand as a mute, five fingers fanning phrases out in all directions. Although he generally remains seated when soloing, standing up only to sing, he still assumes the heraldic posture a vocal-teacher aunt, some seventy years ago, told him would improve his projection—back concave, shoulders straight, elbows akimbo, and head thrown back, with his horn pointed to the rafters and the stage lights shining in his glasses. (This Gabriel imitation, second nature to him, has caused him problems in the recording studios: "The engineers are always yelling at me to point the horn into the microphone. I try, but before I know it, it's up in the air again.")

Watching Cheatham's fingers is an object lesson in proper half-valve follow-through and the nuance it makes possible. But take your eyes off him for a second and you might forget the trumpet is there: the sound emanates so naturally from his person, with so little flourish or strain, that he could be puckering his lips and whistling. A Louis Armstrong medallion given to Cheatham by admirers in Sweden dangles from his trumpet like an amulet, and although he is a less bravura technician, there is something of Armstrong in the way Cheatham, following a string of solos by his sidemen, begins his out-choruses with a thunderclap whole note most other trumpeters would reserve for the climax—this regains the audience's attention right away by prompting speculation about where he'll go from there. Like Armstrong's, his solos tend to be set pieces, the embellishments not varying much from one performance to the next. But hearing him play "Sugar" or "St. Louis Blues" for the umpteenth time can be as rewarding as rereading Raymond Chandler: the rightness of detail continues to enthrall, even when the plot holds no surprise.

Cheatham has an easy stage manner, surprising for a performer relegated to the shadows for most of his career. As prelude to "Miss Brown to You," for example, he might tell the audience how much he loved Billie Holiday, and how he unsuccessfully tried to "hit on her" in 1944, when he backed her on a date for Commodore. "I wrote that in a letter to a British jazz critic once. And you know, he wrote me back and said [Mayfair accent] 'Thank you ever so much for the information, Mr. Cheatham. But there's something I don't understand. If you loved her so much, why did you want to hit her?'"

His rhythm section is wonderfully bracing. The bassist Al Hall is a dead ringer for Harvey Pekar's sagacious co-worker Mr. Boats, and his squat impassivity makes him look several years Cheatham's senior although he's actually a decade younger. But his choice of chords is faultless, as is his slablike beat. Jackie Williams is so self-controlled a drummer that even his exuberant rim shots seem tidy. The pianist Chuck Folds's intelligent lyricism, rooted in James P. Johnson's three-minute stride rhapsodies and Jess Stacy's musings on "Sing Sing Sing," comes to the surface on the behind-tempo preludes that Cheatham favors on the verses to ballads.

Cheatham usually devotes the second of his three sets to requests, which ensures audience participation but limits him to the overfamiliar. Left to his own devices, he'll come up with delightful obscurities, like "Peggy" or "It's the Little Things That Mean So Much"—numbers he sings beautifully, in a small, frayed, conversational voice that betrays little affectation, even though he rolls his 'r's more flamboyantly than anyone this side of Mabel Mercer. "I'm playing all the songs I've learned in the last seventy-five years," he told me when I complimented him on his repertoire. "In the last few years, I've also mem- orized forty or fifty lyrics, because I decided, all of a sudden, that I wanted to sing. Vocalizing gives me a break from having to blow my brains out all the time. It gives me a rest, and I'm entitled."

At this point, Cheatham works as frequently as he wants, although he's careful not to overdo it. He lives uptown with his third wife, Amanda, a retired domestic whom he met during a tour of Argentina with Perez Prado in the fifties (his second wife was Ornette Coleman's aunt). "I play because I want to," he told me, surprising me with the information that he still accepts work as a jazz soloist with society bands. "I could live on my social security. My children are grown with children of their own, and I know how to budget myself. Fresh greens and a couple cans of beans, and I'm happy. I don't need steak. In fact, the doctor told me to cut out red meat. I'm on a fat-free diet. I got all my sickness out of the way when I was younger, except for old folks' sickness, which I guess can't be avoided. I'm having dental problems like I never did before. Your teeth get brittle as you get older, and they move around more. I've had several hernia operations, and the doctors say my gallbladder should be removed, but it isn't really bothering me. They haven't told me to give up cigars yet," he smiled, relighting one. "I've never liked whiskey, so a good cigar is my only vice."

In addition to his weekly gig at Sweet Basil, Cheatham participates in Jack Kleinsinger's frequent Highlights in Jazz concerts at NYU. *Echoes of Harlem*

(Stash), recorded in 1985 but just released, can be recommended for Cheatham's sly take on "Sweet Lorraine" and his evocation of Whetsol on "Mood Indigo," though the rest of the album is a uneven jam, also featuring the veteran tenor saxophonist George Kelly and the callow alto saxophonist and clarinetist Joey Cavaseno. Cheatham's best recent albums, which range from merely excellent to electrifying, are on small Canadian labels: *Doc and Sammy* (Sackville 3013) and *Black Beauty* (Sackville 3029), both duets with the pianist Sammy Price; and *It's a Good Life!* (Parkwood 101), featuring the Sweet Basil quartet, and *The Fabulous Doc Cheatham* (Parkwood 104), with the pianist Dick Wellstood, the prototype for up-to-date anachronisms like Folds. Cheatham is also heard to good advantage on *The Authentic Art Hodes Rhythm Section* (Parkwood 106), a tribute to Armstrong, Bessie Smith, and Ethel Waters, with pianist Hodes and the singer Carrie Smith. Cheatham has also recorded two albums for the banjoist Eddy Davis's New York Jazz label. *Too Marvelous for Words* (J-003) gives him another chance to play lead trumpet, with Davis's Hot Jazz Orchestra of New York. So does *I've Got a Crush on You* (J-001); but the highlight of this album is the side of cozy duets by Cheatham and the talented young guitarist Howard Alden. Both New York Jazz albums feature Cheatham talking about his early career—invaluable oral history.

My favorite among Cheatham's recent efforts is *Black Beauty*, the sort of album that makes you feel, if only momentarily, lucky to be alive and proud to be an American. Subtitled "A Salute to Black American Songwriters," *Black Beauty* provides, in its modest way, as accurate an accounting as we're ever likely to get of the herbs and spices black Americans dropped into the melting pot when no one was looking, and the broth they ladled out. There is no trace here of the medicinal aftertaste that often makes inquiry into jazz and black pop traditions such a bitter pill to swallow—no facile litany of heroes and martyrs. The material chosen by the Canadian producers John Norris and Bill Smith is unhackneyed, much of it reaching back beyond jazz to vaudeville. Echoes of cakewalks, reels, rags, Broadway, the church, and the blues fill both the melodies and the variations that Cheatham and Price wring from them, and the story of early jazz emerges as one of the incidental details. The songwriters represented are Shelton Brooks ("Some of These Days"), Turner Layton ("After You've Gone"), W. C. Handy ("Memphis Blues"), Eubie Blake and Noble Sisle ("Love Will Find a Way"), Donald Heywood and Will Marion Cook ("I'm Coming, Virginia"), James P. Johnson ("Louisiana" and "Travelin' All Alone"), Fats Waller ("Squeeze Me" and "I've Got a Feeling I'm Falling"), and Louis Armstrong ("Someday You'll Be Sorry").

Armstrong is symbolically present in another way—trumpet and piano duets in the vintage style favored by Cheatham and Price inevitably recall Armstrong's duets with Earl Hines. *Black Beauty* is more reflective in mood, however; Cheatham and Price are hardly weather birds eager to brave the winds of change, as Armstrong and Hines were almost sixty years ago. Still, with their vigor, Cheatham and Price nicely refute the romantic notion that jazz is exclusively a young man's forum. Perseverance has only recently come to be recognized and valued in jazz performers, and these men—hidden away in big bands and studio orchestras—are making up for lost time. A more limited player than Cheatham, Price relies on a stockpile of boogie-woogie and barrelhouse phrases, but he delivers them with infectious bounce and replenishes them with lacy Tatumesque filligree. He rides Cheatham with droll pungency, though he doesn't always get in the last word—the second time he quotes Bessie Smith's "Gimme a Pigfoot" in "After You've Gone," Cheatham responds with the refrain from "You're Getting to Be a Habit with Me."

Some truculent modernists could learn a thing or two from these men. When Cheatham shows off his still-gorgeous upper register, he's not demanding adoration, the way a Freddie Hubbard might—he's generously inviting us to partake of the pleasure he himself derives from his skills. The bothersome question of whether a jazz musician is an artist or an entertainer seems never to have vexed Cheatham or Price. They're artists, all right, but they realized early on that there would be no audience for their art unless they also entertained. And unlike the pandering of too many of their successors, their decision to entertain doesn't mean they assume the worst about their audience.

Black Beauty approaches perfection. Still, the ideal way to hear Doc Cheatham is at a Sweet Basil matinee. There's no better nourishment for a Sunday afternoon.

(*April*, 1987)

THEY CAN'T
TAKE THAT AWAY

It was a popular sentiment of the 1940s, best articulated by Brew Moore: Anyone who didn't play tenor saxophone like Lester Young was playing it wrong. Yet Young himself might have flunked that test in the fifties. I pinpoint the last decade of Young's life because, skeptical though I may be of Lewis Porter's absolute faith in musicological empiricism, I agree with Porter's thesis (in *Lester Young*, Twayne, 1985) that the standard method of dividing Young's career into pre- and postwar halves misses the point.

There were three Lester Youngs, not two: Young acknowledged as much in the last interview he ever gave, when he rationalized, "I've developed my saxophone . . . to make it sound just like an alto, make it sound like a tenor, make it sound like a *bass*. . . ." (Another way of putting it was that his "voice" dropped as he grew older, much like that of a singer.) The first Lester Young was the (now) universally admired Count Basie sidemen who revelled in the paradox of improvisational confidentiality amid big band ruckus. The second was a composite of the stockade casualty who should never have been inducted into the service in the first place, and the lone-wolf soloist many thought should never have left Basie—it was this Lester who, whatever his problems, was self-possessed enough to create lucid improvisations like those on "Just You, Just Me" (1943); "After Theater Jump," "Three Little Words," "Indiana," and "Blue Lester" (1944); "D. B. Blues" (1945); "Back to the Land" and "Somebody Loves Me" (1946); "Jumpin' with Symphony Sid" (1947), and "These Foolish Things" (1944, 1945, or 1952—take your pick). The final Lester Young, who continues to elicit morbid fascination while eluding critical consensus, was the self-condemned man of the LP era, chain-smoking and yawning for the guillotine.

The transcript of Young's 1945 court martial is now on public record, thanks to the Freedom of Information Act and the diligence of John McDonough and Ann Holler (the annotators of Times-Life's Lester Young box). But given the casual acts of violence that are the rule behind bars, that neither victims nor perpetrators are likely to talk about even off the record, the daily assaults on self-esteem that Young must have endured in an army detention barracks *after* being judged guilty will always be open to conjecture. Though Young's decline was too gradual to be solely the aftermath of stockade trauma, his incarceration was surely a factor—along with gin and mary jane (to say nothing of nicotine and poor nutrition) and the harmonic veer of the bebop revolution, which overnight transmogrified the hep-cat perception of melody-man Lester from visionary to anachronism. For whatever combination of reasons, Young's confidence seemed diminished after 1952. His sequences were full of faltering steps where there used to be gracious leaps, and his tone became woozy and waterlogged, creating the illusion of a slower, more phlegmatic vibrato. While it is easy enough for revisionists like Porter to invoke Young's famous quip about not wanting to be "a repeater pencil," the recorded evidence suggests that

Young would have been unable to replicate his early solos with Basie, Billie Holiday, and the Kansas City Six even if he'd wanted to.

It's mortifying to listen to some of Young's later solos—in particular, "They Can't Take That Away from Me," recorded on February 8, 1958, just a year before his death, and included on *Laughin' to Keep from Cryin'* (Verve UMV-2694). As one of only two clarinet features from his last decade (the other was "Flic," a middling blues from six months earlier), "They Can't Take That Away from Me" would seem to be one of the choice items in Young's latter discography, if taken at face value—a belated companion piece to "Blue and Sentimental" with Basie and the eloquent "I Want a Little Girl" with the Kansas City Six! But Lester's embouchure is so weak he can barely frame the notes. Trumpeters Roy Eldridge and Sweet Edison solo gamely in what must have been worrisome circumstances, and the pianist Hank Jones valiantly reiterates the chords, as though trying to prop up Lester's dead weight. But it's no use—Jones, Eldridge, Edison, the guitarist Herb Ellis, the bassist George Duvivier, and the drummer Mickey Sheen are like pallbearers carrying Lester to his final resting place. In all fairness, Young was struggling with an instrument he seldom played, and his choruses do pick up momentum as they go along, which suggests that producer Norman Granz should have called for another take. But after hearing countless tenor solos from the same date and earlier, on which Lester runs short of breath or seems to lose interest halfway through, who can swear that another take would have yielded anything better? The first time I heard this death rattle of a solo by an immortal musician I adore, realizing with a shock that he was only forty-eight when it was recorded (seven years younger than Sonny Rollins is now), I felt the same mix of betrayal and despair that was expressed by one of the characters in Jean-Luc Godard's *Masculin-Féminin*: "We often went to the movies . . . the screen lit up and we trembled . . . but more often than not we were disappointed. The pictures were dated, they flickered. And Marilyn Monroe had aged terribly. It made us sad. . . . This wasn't the film we'd dreamed of. This wasn't the film that each of us had carried within himself, the film we wanted to make, or more secretly, no doubt, wanted to live."

So why do I listen? Why do I continue to scout import and secondhand bins for Lester Young on Verve? Maybe for the unexpected epiphanies, the dragging of emotional depths only hinted at in his halcyon years, the sudden and evanescent recoveries of form that he was capable of from chorus to chorus to the bitter end—characteristic less of a death wish than a stubborn will to go on living for no good reason at all. The occasional triumphs of his later years, far from being mere epilogue, amount to a substantial body of work all by themselves, offering insights that are no less valuable for being less sanguine than those he shared in his prime.

Some knowledgeable listeners find value even in Young's most desultory later work. In his Young biography (in Hippocrene's Jazz Masters Series), the British saxophonist and critic Dave Gelly hails Young's theme statement and opening chorus on "They Can't Take That Away from Me" as "the last indisputably great statement of his recorded career. The wraith-like tone is so fragile that notes occasionally fail to make it out of the instrument, but as a paraphrase of the theme this solo is a masterpiece. . . . The other players on the session were in far better shape than Lester, but this last, plaintive gasp of genius drowns all their robust craftsmanship." I'm not quoting Gelly to mock his subjectivism; my affection for "Almost Like Being in Love," recorded during a disastrous French tour on March 4, 1959 (*Lester Young in Paris*, Verve MGV-8378), less than two weeks before Young's death, leaves me equally vulnerable to ridicule.

But two 1956 dates with Teddy Wilson need no apology from me or anybody else: one, a quartet with the bassist Gene Ramey and the drummer Jo Jones (included on *Pres and Teddy and Oscar*, Verve VE2-2502); the other, an all-star affair adding Roy Eldridge, Vic Dickenson, and Freddie Green (*Jazz Giants '56*, Verve UMV-2511). This is Lester as we wish to remember him, orbiting melodies and eliding chords, rubbernecking around bar lines and playing catch-up with the beat, beginning his solos in a mossy lower register before soaring like a butterfly into his final choruses, with Jo Jones's switch from brushes to sticks underlining the metamorphosis. If I had to choose just one Lester Young performance for a desert island, is would probably be "This Year's Kisses" from *Jazz Giants '56*, for the unimpeachable architecture of his solo and the camaraderie among him, Wilson, Eldridge, and Dickenson. In lieu of "Tickle Toe" or "Taxi War Dance," others might pick "Prez Returns" from the Wilson quartet date, a series of undulating riff variations that are perfect miniatures of the ones Young patented with Basie twenty years earlier. "Prez Returns" is the most widely celebrated of the many blues that Young recorded in the fifties, but it's hardly the only great one. On a 1955 date with Sweets Edison and Oscar Peterson (*Mean to Me*, Verve VE2-2538), there is a slow, teasing number called "Red Boy Blues" that is full of the rending moans that Young perfected in the late forties, and that spawned the funky tenors just as surely as his thirties lilt spawned the Four Brothers. (Black and white saxophonists responded to different qualities in Lester.) Of course, Young was one of those players who turned everything he touched into the blues, particularly in later years, and his 1953 interpretation of "Tenderly" (*Lester's Here*, Verve

MGV-8161) is remarkable for the way he distances himself from the written melody in the process of arriving at a new one at once earthier and more abstract. And no survey of late Lester Young is complete without mention of Pablo's four-recorded series, *Lester Young in Washington* (2308-219; 2308-225; 2308-228; 2308-230), which lends credibility to eyewitness claims that he was in better form in nightclubs than in the studio in later years, and which unearthed at least two great performances: a "These Foolish Things" arguably superior to the ones on Savoy, Aladdin, and Norgran; and a "Tea for Two" linking him to Sonny Rollins as a wizard of tonal embellishment, rhythmic displacement, and thematic exposition—ironic, because Rollins has always favored a big, rolling, Hawkins-like tone, and thus would seem to be the antithesis of Young.

The mention of Coleman Hawkins raises futile speculation; what if Young had aged as gracefully as Hawkins did, and what if he had monitored ongoing progress in jazz as diligently? Though it's impossible to imagine Lester recording with Rollins, Thelonious Monk, Orchestra USA, or Max Roach, Abbey Lincoln, and Michael Olatunji, as Hawkins did, he might have been comfortable with Miles Davis, George Russell, the Jimmy Giuffre 3, or the Modern Jazz Quartet. Given his reported listening habits and his documented rapport with Billie Holiday, Jimmy Rushing, Helen Humes, Billy Eckstine, and Una May Carlisle, it's tempting to imagine him blowing obbligatos behind Frank Sinatra, Kay Starr, Jeri Southern, or his beloved Mills Brothers. It's equally tempting for the ear to substitute Lester for his disciple Paul Quinichette on the latter's 1957 brushes with John Coltrane. That Norman Granz never acquiesced to Young's desire to record with strings is probably just as well. But suppose Granz had encouraged more elaborate projects, and suppose the arranger had been Gil Evans?

An overriding problem with Young's fifties albums is that apart from marginal improvements in recording quality and unaccountable fluctuations in Lester's tone, it's impossible to date them without access to a discography—even when healthy, Lester was in a rut, and Granz dug him in deeper. Still, it's time for PolyGram to give Lester the royal treatment accorded to Charlie Parker and Billie Holiday, and reissue his entire output for Granz in chronological order in a deluxe box set. With the third and final Lester Young, you want all of it, the bad as well as the good, because it's strictly a matter of opinion which is which.

(July, 1986)

POSTSCRIPT

In 1987, PolyGram reissued the 1952 LP *The President Plays with the Oscar Peterson Trio* on compact disc (Verve 831 670-2), with five previously unissued tracks, including two takes of Young singing(!) "Two to Tango." Ironically, the man who insisted that you had to know a song's lyrics in order to turn in a credible interpretation here forgets the words. No matter—he recovers by substituting his own risqué lyrics and tossing off the most beatific scat choruses this side of Louis Armstrong (to whom he bears an uncanny vocal resemblance). The mood of this performance is lighthearted, and so is Young's between-takes banter with Granz—proof that the stereotype of Young as terminally out-of-it and despondent in his final decade is an oversimplification.

TALKING WITH MILES

"You talk to me like I was just born with a trumpet in my mouth and that was it. Man, I *studied*," scolded Miles Davis, the only musician I ever phoned on behalf of *The Philadelphia Inquirer* who sounded disappointed when he found out it wasn't *The National Enquirer*. "I'm a musician, right?" he asked rhetorically, in that raspy whisper. "I deal in styles. I play different styles, you understand? When I'm in a Chinese restaurant, I eat Chinese food. When I'm in a Japanese restaurant, I eat Japanese. When it's a ballad, I play a ballad. If it's funky, I do that. The same way I treated *Sketches of Spain*, I treated *Tutu*. I treat all music the same. Respect it, and do your best. If you can't do it, don't even try. I don't *try* anything, I do it."

Davis, enduring the interview in his Central Park apartment, said he'd just come back from Los Angeles, where he spent the better part of a month laying down tracks for a new album, his second for Warner Brothers after thirty years with CBS. Although he'd reveal only that it would contain "more music," his current listening preferences—"everybody on the black stations: Cameo, Prince, Jimmy Jam, Colonel Abrams, the Time, Michael, Janet, everybody," he said, turning the radio up full blast and holding it against the phone, becoming annoyed when I couldn't make out what was playing—indicate that his next LP will be another attempt to get on the good foot, just like last year's *Tutu* (Warner Brothers 25490-1). *

That'll strike some people as good news. After all, *Tutu* won Davis a Grammy for the best performance by a jazz soloist. But here's what puzzles many of his longtime admirers, myself included. If, as Davis seems to believe, music is music, and staying contemporary isn't the issue, why doesn't he surprise everyone with an acoustic jazz album?

"You know people who think that? I don't think they think that. You don't know the public. The public is a lot hipper than you think they are. You don't know what the fuck you're talking about, Francis." Miles is notorious for not granting many interviews and not being especially cooperative when he does. Yet like many public figures who have been interviewed thousands of times over the years, he's mastered the trick of remembering and periodically using his interrogator's name—the difference is that he uses your name when he's cussing you out. This should add insult to injury, but it's actually pretty charming. It does make exact quotation in a paper like the *Inquirer* difficult, however, and yet to bowdlerize him does him an obvious disservice, because his profanity gives his speech its rhythmic thrust, much the way Philly Joe Jones's rim shots once propelled his solos. Once an interviewer demonstrates that he can withstand Davis's rancor, he becomes surprisingly open, answering the questions he thinks he *should* have been asked.

Davis is one of the half dozen or so most important figures in jazz history, and the only musician active in the bop era whose work still incites controversy. Only Duke Ellington has had as far-ranging and as long-lasting an influence on other musicians, and only Louis Armstrong, Charlie Parker, and Ornette Coleman have altered the course of jazz more abruptly. In the forties and fifties, when bop and cool were still regarded as antithetical, Davis played a large role in the gestation of each, going straight from an apprenticeship with Charlie Parker to collaborations with Gil Evans, Lee Konitz, and Gerry Mulligan. *Kind of Blue*, recorded in 1959, with John Coltrane, Cannonball Adderley, and Bill Evans, anticipated nearly every significant development of

---

* Not exactly. Davis's next release was *Music from Siesta* (Warner Brothers 25655-1), the soundtrack from a pretentious 1987 film directed by Mary Lambert and starring Ellen Barken. If this was the album that Davis had just come back from recording when I spoke with him, no wonder he mentioned *Sketches of Spain*. In addition to boasting the most haunting solos that Davis had recorded since 1983's *Star People*, the soundtrack was also fascinating as the composer and producer Marcus Miller's homage to Gil Evans—a synthetic *Sketches of Spain*, as it were. This was ironic, given that Miller was prematurely praised by some critics for evoking Evans on the deplorable *Tutu*. *Music from Siesta* had more orchestral verisimilitude, though it sounded like a stunt.

the next decade, including free form, modal impressionism, and soul. The harmonically spacious, metrically suspended music he played with Wayne Shorter, Herbie Hancock, Ron Carter, and Tony Williams in reaction to Coleman in the mid-sixties remains the dominant style of jazz two decades later, although no one has yet put a satisfactory label on it. In 1969, Davis became an avatar of jazz-rock fusion when he hired the guitarist John McLaughlin and recorded *In a Silent Way* and *Bitches Brew*. Ironically, just as his defection to electric music alienated many of his older fans, there are now younger listeners who mourn Davis's rejection of fusion for a slicker brand of funk.

For twenty-five years, beginning in the late forties, Davis's music was a sure index of progress in jazz, and one cannot attach too much significance to the fact that jazz took a nearly fatal commercial dive during the six-year period (1975–81) that Davis withdrew from recording and live performance. But since his comeback five years ago, Davis has no longer been setting trends, just following them. In attempting to crack urban contemporary radio, he isn't selling out so much as buying in—the baddest horn in jazz gamely trying to impress his badness on younger black listeners for whom a horn is merely a prop for puttin' on the hits. Still, the hit single that Davis needs to establish contemporary relevance continues to elude him. At this point, following his Honda commercials and his guest shots on *Crime Story* and *Miami Vice*, he's in danger of becoming a celebrity, one of those people famous for being famous.

On stage, he remains the most electrifying figure in jazz. He has his best band in years, with the tenor saxophonist Gary Thomas and the percussionist Marilyn Mazur prominently featured. Thomas plays dark, slow-building solos that match Davis's own for lyrical intensity, and Mazur—caged in by her various floor drums and overhanging rhythm tubes on an elevated perch at stage right—is fun to watch crawling around even before she leaps down alongside Miles to shake the bells on her ankles and hips ("Miles' percussion slave," a friend of mine described her). Miles himself has become a kind of roving conductor, walking from musician to musician, describing the rhythms he wants from them with a pump of his shoulders, blowing riffs in their faces and letting them pick it up from there. He still spends a good deal of each set with his back turned to the audience, like he was scolded for doing in the fifties and sixties. The difference now is that he has a microphone attached to his horn that allows him to be heard with his back turned—and, at this point, audiences would be disappointed if he failed to strike this iconic pose. The last time I saw him, he was in an uncharacteristically accommodating mood, even introducing the band members by name (first names only, granted—but that's more than he ever did for Cannonball and Coltrane). Though the show was never boring, I wasn't sure how I felt about it afterwards, and I'm still not. Is Miles admirable, as his apologists would have it, for refusing to rest on his laurels, for keeping up with the latest black musical and sartorial trends? Or is there something pathetic about the sight of a sixty-year-old man in jheri curls and platform heels shaking his ass to a young man's beat? Musicians from earlier generations also prided themselves on keeping up, but they also kept some perspective on what was for kids and what was for keeps. Does Miles know the difference anymore? Does anyone these days?

Tutu, with Davis sulking and attitudinizing (minus his band) above synthesized, assembly-line arrangements of the clock-punching funksters Marcus Miller and George Duke, was noteworthy only for Irving Penn's cover photos, which transformed Davis's handsomely creased face into something like an African mask. Everything else about the album was generic, save for Davis's wraithlike solos, which, though venturing nothing substantive, were instantly identifiable as his, if only by their timbre. Miller's "Splatch" and Duke's "Backyard Ritual" were marginally more palatable than the other tracks, but only because they were bouncier, not necessarily more involving. And if Miller's "Full Nelson" and the cover version of Scritti Politti's "Perfect Way" were slightly unsavory, it was because they found Davis contemptuously alluding to better days, with throwaway snippets of "Fran Dance" and "Half Nelson." As a result of Davis's self-reflexiveness, *Tutu* sounded like nothing else released in 1986—but not from lack of trying. If not for its reggae-*cum*-hip hop afterbeat and slight existential twist, *Tutu* could have been a new release by Herb Alpert or Chuck Mangione.

Which is not to say that Davis no longer wields influence. In fact, "Keep Your Eyes on Me," Alpert's current smash, sounds suspiciously similar to the tracks on *Tutu*, albeit with a hotter rhythm mix (courtesy Jimmy Jam and Terry Lewis, who also produce Janet Jackson) and without the foreboding that rescued *Tutu* from complete frivolity but doomed it as a dance groove.

"If that's what it sounds like to you, that's what it sounds like to you," Davis said, when asked if he heard the similarity. "I like Herb. I don't like that record. I like it when he does *his* thing. He's not black. . . .

"You thinking for me now? You don't know what the fuck I think. Don't put words in my mouth," he warned, when I asked him if his well-publicized hatred for the word "jazz" stems from the false expectations the word creates in the minds of listeners who prefer his music from twenty or thirty years ago. "I don't like the word because the record companies don't push it when it's called that, because white people want to protect their daughters' ass. They think all jazz musicians want to seduce their daughters—and white people have told me that. That word has been worn out and never been used right anyway. It meant, 'Let's get these people to march for us and entertain us the way they did in New Orleans.' Nobody paid attention to the music. It was just the smiling and the shining of teeth. Now they're trying to clean it up, build monuments to jazz and all of that shit, but it's too fucking late, man.

"I went down to Washington for Ray Charles', what do you call it, [Kennedy Center] Lifetime Achievement Award, and, Francis, it was sad. Cicely [Tyson, the actress, his third wife] started crying because all these black kids were there to honor Ray. It was touching, man, but the white people there, politicians and their wives, didn't feel nothing. They didn't know what to feel. I felt sorry for them, damn. They act so sophisticated, and they don't know nothing.

"Come on, what else. I gotta go," Davis said impatiently, abruptly changing mood. He confided that, yes, his long-rumored collaboration with Prince might yet take place, boasting of just having spent a few days with Prince on the West Coast. (Shouldn't Prince be the one boasting?) And, yes, he intended to do more TV—in fact, he has just completed a pilot for a series called *Shake and Bake,* in which he plays private eye Tom Skerritt's wheelchair-bound, ex-gangster sidekick.

Just one more thing, Miles: Do you agree with the people who say that a player's tone is an accurate projection of his character?

"What? You gotta ask a white psychiatrist about that, not me," he said, perhaps sensing that the next question would have been what his sound—the juxtaposition of his vulnerable trumpet style and his curt, self-aggrandizing manner—revealed about him.

I wanted to ask him if he still listened to singers like Frank Sinatra and Tony Bennett, as he reportedly used to, or if he had rejected them for not being "contemporary" enough. But it was clear that he wasn't going to indulge any more of my foolish questions. In exchanging goodbyes, I remarked that, in light of the autobiography he was working on with the writer Quincy Troupe, I was surprised that he was still granting interviews. Wouldn't most musicians in that situation figure they'd save it for the book?

"Including me. But I don't mind just talking, you know? But a lot of the shit you ask me, I just don't know. You ask me questions an anaylist would ask."

In that case, maybe I should send him a bill.

"That's right," he said, laughing. "I'll pay you back when I play." Then he hung up.

(May, 1987)

WEST COAST GHOST

The hottest new name in jazz is that of a fifty-three-year-old alto saxophonist who sacrificed his youth to heroin. Frank Morgan was one of a legion of musicians doomed to addiction by their idolatry of Charlie Parker, whose habit was as mammoth and as legendary as his genius. "I think it was 1952, when Bird came back to L.A. for an engagement," Morgan says of breaking the news to Parker that he was "a member of the club." "It broke his heart," Morgan says. "He said, 'I thought you would be the one who had sense enough to be able to look at what it's done to me.'" The sermon ended when Morgan flaunted the half ounce of heroin and the half ounce of cocaine he'd just scored. "After we got high, he talked to me about dying. In a tragic sense, I think Bird felt he could set a better example by dying."

If so, his disciples failed to heed the warning. Upon learning of Parker's death (at the age of thirty-four, in 1955), Morgan and several other participants at a Los Angeles jam session mourned him by shooting up and playing "Don't Blame Me." A few months later, *Frank Morgan* (GNP Crescendo GNPS-9041), Morgan's first album as a leader, was released, with hard-sell liner copy nominating him as the new Charlie Parker. By that point, Morgan was well on his way to a hundred-dollar-a-day habit, which he eventually supported by forging checks and fencing stolen property. First arrested in 1953, Morgan spent the better part of the next thirty-two years in California county jails and state penitentiaries, and the few critics and record collectors who remembered him assumed that he was dead.

I spoke with Morgan this spring, during his second visit to New York, a city with good associations for him. The day before his first trip East, last December, a California court lifted his parole, which was scheduled to last until 1988 and would have prevented him from accepting lucrative offers to perform in Europe this summer. During that first week-long stay, he recorded a live album at the Village Vanguard and was in constant demand for interviews for the first time in his career.

He was in perpetual motion on his return trip; going to a methadone clinic early every morning, recording an album with the pianist McCoy Tyner over the course of three long afternoons, and performing three sets a night at the Vanguard, usually not returning to his hotel until 4:00 P.M. There was a near-crisis on opening night at the Vanguard. Morgan fired a veteran New York pianist who he thought was feeding him the wrong chords. This distressed Morgan, first of all because the pianist had recorded with Charlie Parker in the 1950s, and second because he worried that the New York musicians whose favor he was courting would be up in arms if an outsider found fault with one of their own. But the musicians Morgan talked to, including Cecil Taylor, assured him that they sympathized, and he basked in their approval. During his stay, he met with the *New Yorker* staff writer George W. S. Trow, who sought his collaboration on a musical drama. Trow says that the play, which depicts some incidents from Morgan's life, as well as from his own, was partly inspired by the "nobility" he perceives in Morgan, as a musician and as a man. Morgan will have both a musical and a speaking role in the play, which will have its premiere in New York next month. *

It was difficult, as I shared a taxi from the recording studio with Morgan and his entourage, not to see the city through his eyes. With rain on the way, the skyline looked even more oppressive than usual, to anyone used to looking at it. But this was the skyline Morgan had imagined from countless prison cells—a skyline with his star glittering on the horizon.

A free man at last, with three highly rated albums to his credit since 1985, Morgan is obviously good copy—which, one suspects, is why he has been interviewed for *Newsweek* and *CBS Sunday Morning. People* magazine even took him back to San Quentin for a photo shoot. The mass media have no great love for jazz, but they do love a success story—especially (now) one whose moral can be boiled down to the sanctimonious slogan "Just Say No."

But Morgan has also attracted the attention of fellow musicians, who have seen far too many redeemed junkies rise from the ashes to work up much enthusiasm for another one, unless he has something else going for him. Players who emerge from oblivion to make waves are generally either prophets (like Ornette Coleman, when he introduced free improvisation in 1959) or

---

* *Prison-Made Tuxedos* (the title refers to Morgan's experience with the San Quentin band, which would perform every Saturday, in tuxedos made by fellow prisoners, for visitors making the grand tour of the prison) opened at St. Clement's Church in New York in November, 1987. Morgan was responsible for most of the play's poignant moments—reminiscing about how a prison sentence prevented him from attending his grandmother's funeral, for example, or about how he was attracted to Charlie Parker's erudition as well as his music. Morgan also played, and in jazz parlance, his music was saying something. In contrast, Trow was merely *trying* to say, like Benjy in *The Sound and the Fury*. Morgan's part of the show was nakedly autobiographical, but Trow hid behind satirical convention, never coming clean. The uneasy juxtaposition of their lives convinced you that Trow is one of those guilty white intellectuals who thinks of inhibition as White Man's Disease.

anachronisms (like Bunk Johnson, rediscovered in Louisiana two decades earlier). Morgan is neither. He plays bebop, which is still the most prevalent form of modern jazz although it is now over forty years old and has been subject to endless permutations. But he plays it with an urgent authenticity that inspires fantasies of the Royal Roost and Minton's in the era when Parker himself held sway, before bebop splintered into hard bop, "cool" jazz, and free bop, and finally devolved into cliché—the price it paid for becoming the most prevalent form of modern jazz.

In one sense, Morgan is a West Coast Ghost—the title of a Charles Mingus composition, and Mingus's apt designation for those California-bred musicians like himself who chose New Yorkers like Parker and Dizzy Gillespie as their role models. Because jazz history is as prone to oversimplification as any other branch, jazz on the West Coast in the forties and fifties is supposed to have been mentholated and supernal—*cool* in polarity to bebop's scalding heat. Morgan's fire is a necessary reminder that bebop became a bicoastal phenomenon during the sixteen eventual months that Parker spent in California, beginning with an engagement at Billy Berg's in Hollywood in December 1945. Bop was also biracial, of course; as was cool, which was essentially a bop offshoot. But because the overwhelming majority of records issued by West Coast labels during the period featured white musicians with temperate styles, listeners in other parts of the country were given an incomplete picture of California jazz. "We weren't part of that cool scene at all," Morgan says of the Los Angeles bebop underground, which included the trumpeter Howard McGhee, the alto saxophonist Sonny Criss, the tenor saxophonists Dexter Gordon and Wardell Gray, the pianist Hampton Hawes, and others who have been forgotten. "It was a racial squeeze. Only a few of the black musicians on the West Coast were being recorded."

But despite his loyalty to Parker, Morgan's penchant for improvised counterpoint when working with another horn suggests that cool's niceties haven't been lost on him, just as the overblown notes and scalar runs that dot his up-tempo solos suggest a willingness to use techniques associated with free jazz to give bebop an unexpected modern spin. Morgan was still using heroin and wanted for parole violations when he recorded the understandably furtive *Easy Living* (Contemporary C-14013), his first album as a leader in thirty years, in June 1985. *Lament* (Contemporary C-14021), recorded almost a year later, after Morgan turned himself in and served six months, was the album that announced his resurrection. *Lament*'s ballads showed that Morgan shared Parker's speechlike delivery, his courtly approach to melody, and his knack for transforming pop songs into the blues. In other words, it was clear that Morgan was blessed with what might be called "the common touch," a way of appealing to the unschooled listener on the most basic musical level even while improvising lines of baffling harmonic complexity. His passionate reading of the folk singer Buffy Sainte-Marie's "Until It's Time for You to Go" was a vivid illustration of the power of ardor to rehabilitate treacle (Morgan's two versions of "Theme from Love Story"—with George Cables on *Double Image,* Contemporary C-14035; and with the McCoy Tyner Trio on *Major Changes* Contemporary C-14039—are even better examples of this), and Morgan's nimble choruses on Wayne Shorter's moody "Ava Maria" demonstrated his ease with non-bebop repertoire.

Despite its more conservative material, *Bebop Lives!* (Contemporary C-14026), recorded Morgan's Village Vanguard debut in 1986, is the album that captures him at his peak. The flugelhornist Johnny Coles's sidelong phrasing contrasts nicely with Morgan's more frontal attack, and they banter good-naturedly in stating the theme of Thelonious Monk's "Well, You Needn't." Morgan proves that he's capable of running the chords with the best of them on Jackie McLean's "Little Melonae" and a version of Cole Porter's "What Is This Thing Called Love" jammed with allusions to Tadd Dameron's "Hot House," one of the song's more ingenious bebop derivatives. But Morgan's unaccompanied introduction to Jerome Kern's "All the Things You Are," with its broken cadences and phantom glisses, is finally the most convincing display of his mastery. His solitary wail offers needed reassurance that bebop still holds potential for soul-searching introspection, not just breast-beating exhibitionism.

Morgan, who was born in Minneapolis in 1933, is a second-generation musician. His father is Stanley Morgan, a guitarist who recorded with Harlan Leonard and the Rockets in 1940 and now leads one of several different groups performing as the Ink Spots. Stanley Morgan groomed his son as a musician even before he was born. "My mother recently told me that when she was pregnant with me, my father would stand behind her and reach around her with his guitar, leaning the back of it up against her stomach as he strummed, so that I would pick up the vibrations." Because his parents were frequently on the road, Morgan lived with his grandmother in Milwaukee from the age of six to fourteen, when she put him on a train for Los Angeles, where his father had opened a club. "She caught me with a joint, and felt the time had come for a father's guidance." In Los Angeles, Morgan studied at Jefferson High School with Samuel Browne, a music teacher whose other students over the years have included Don Cherry, Art Farmer, Dexter Gordon, and Wardell Gray. At fifteen, he won a TV talent-show contest, and got to record "Over the Rainbow" with the Freddy Martin Orchestra as a prize (the arranger was Ray

Coniff, and the singer was Merv Griffin). By 1952, he was good enough to win a seat in Lionel Hampton's saxophone section (like most musicians who've been with Hampton, he complains that the vibraphonist paid badly, and added insult to injury by always bumming cigarettes from him). His first arrest cost him the opportunity to join the original edition of the Max Roach–Clifford Brown Quintet.

According to Morgan, it wasn't just his addiction that was responsible for his long pattern of recidivism. He was a privileged character behind bars, admired by fellow inmates for his musical prowess and his adherence to the underworld code (he once refused to lead authorities to his higher-ups in a forgery ring, despite being offered leniency if he would do so). "I went to great lengths to be a stand-up criminal that other criminals would admire. I was charged with a hundred thousand dollars' worth of forgeries, and I didn't have a dime." After his refusal to cooperate, he was sent to San Quentin, where he had everything he felt he needed, including plentiful supplies of heroin and ample opportunity to practice his horn and perform regularly with the prison band. "That was good for me, because I was the kind of junkie who would forget all about music when I was out on the street trying to score." Had he so desired, he could have even had his choice of male sexual partners. "The night I arrived in San Quentin, I had fifteen or twenty 'ladies' to choose from. Not real ladies—that's the prison vernacular for them. But that wasn't my thing, and one of the first things you're warned about in prison is to stay clear of that if you want to live. Sex is what most of the killings are about, not drugs. People actually consider themselves married, and the jealousy is intense.

"I had everything I wanted except for freedom, privacy, dignity, self-esteem—everything I value now. In San Quentin, I had a better rhythm section than I had at the Vanguard last night. Maybe not better man-for-man, but, for me, prowess isn't as important as rapport. But I couldn't fire a musician, like I did last night. I'd have been stuck with him until he got paroled, or until another stupid musician who played better went and got himself locked up. In prison now, there might be good rock rhythm sections, but they couldn't give *me* the guys to play with, and even if they could, I wouldn't want to go back. The prison population has changed, or maybe I've changed. The guys I just left in prison, I don't want to be around those motherfuckers. I'm not like them."

Ten years ago, the alto saxophonist Art Pepper—another West Coast Ghost, and a former bandmate of Morgan's at San Quentin—arrived in New York with a story much like Morgan's, but with more squalid and frightening details. Pepper also recorded during a triumphant engagement at the Village Vanguard, but later admitted (in his 1979 autobiography, *Straight Life*, co-authored by his wife, Laurie Pepper) to shooting so much cocaine that he practically had to be carried to the club on the final night. Although Pepper's solos took on awesome power in the five years between his Vanguard debut and his death in 1982, suspicion lingers that he never successfully kicked his habit.

A cynic might observe that Morgan, having blown his chance to become the new Charlie Parker, might instead become the new Art Pepper. (Contemporary is a subsidiary of Fantasy Records, as is Galaxy, Pepper's final label.) Morgan is clean, but will he be able to stay that way? He has a supportive lover in Rosalinda Kolb, a painter and photographer. But he and Kolb have been a couple since the late seventies, and she admits that she wasn't able to save him then: "He would lie to me, and say that he wasn't using, and there would be an element of self-deception on my part. To be able to stay with Frank, I had to look the other way a lot. But I knew what was happening, and it made me sick to watch him."

I'm told by those who have worked with him that Morgan is capable of losing his temper if that's what it takes to get his own way. Still, he's visibly milder in temperament than Pepper, whose criminal exploits seemed to allay his early feelings of inauthenticity as a white musician in a field dominated by blacks. I suspect that Morgan relishes the newfound attention he's receiving too much to risk losing it by reverting to his old ways. Success, now that he's had a small taste of it, might prove to be the most addictive high of all. "Before, I always had excuses. I remember when Ornette Coleman came on the scene, I lost hope. I said, 'I'm not going to play this.' Same thing when Ronald Reagan was elected president. I milked that one pretty good. When I was offered a recording contract a day after I got out of prison, on April 2, 1985, it scared the shit out of me. I think it contributed to my using again. But when I walked into the studio [two months later, to record *Easy Living*] and played with Cedar Walton and Billy Higgins [a pianist and a drummer, respectively, both much better established than Morgan at the time], and found out I could do it—that, in fact, it maybe wasn't even the ideal rhythm section for what I wanted to do—everything changed. I no longer had an excuse to fail."

(November, 1987)

AN AMERICAN
IN PARIS

I once heard a radio interviewer introduce the soprano saxophonist Steve Lacy, who was born in New York in 1934 and has lived in Paris since 1970, as a musician who has played every jazz style from the most archaic to the most advanced. Yes, but not exactly. It's true that Lacy started off as a dixieland revivalist in the mid-1950s, only to wind up in the front ranks of the jazz avant-garde before the decade was over. But his evolution hardly parallels that of jazz, because he skipped a crucial step in between—and this is part of what makes him such an original. He was briefly a sideman with Thelonious Monk in 1960, and has since distinguished himself as Monk's most steadfast interpreter. But unless you count Monk's music as bop by virtue of its contemporaneous origin (there is no other reason for doing so) Lacy has never really played bop, the dominant style now as in the 1950s, when Lacy began his career.

Lacy has always gone his own way. Even his choice of instrument contributes to his individuality. Sidney Bechet, the first great jazz soloist (he preceded Louis Armstrong), favored soprano saxophone over clarinet, and the alto saxophonist Johnny Hodges frequently played soprano with the Duke Ellington Orchestra in the thirties and forties. But until Lacy recorded with the pianist Cecil Taylor in 1955, the horn had no credibility in modern jazz. Even now that many alto and tenor saxophonists double on soprano, Lacy is one of very few musicians playing it exclusively, and its popularity has little to do with him, except indirectly: he piqued the interest of the late John Coltrane, whose 1960 recording of "My Favorite Things" made him the dominant influence on the horn (though Coltrane never achieved the fluency on it that he had on tenor saxophone, his main instrument).

Soprano is the most antagonistic member of the saxophone family: the most difficult to tame and to keep in tune. "It took a certain combination of ignorance and courage," Lacy told me last spring, when he was back in the States for an engagement at Sweet Basil in New York. "I didn't know it was hard to play. I didn't know it wasn't being used, and I didn't know I wouldn't get gigs on it." The trombonist and critic Michael Zwerin, Lacy's friend and fellow expatriate in France, once said that Lacy now gives the impression of playing soprano *despite* its popularity; and despite Lacy's protestations that it's not he but the horn that's perverse, there's an element of truth in Zwerin's humorous observation. In contrast to the snake-charmer drone of Coltrane's imitators, Lacy's tonal flexibility makes him sound as individual now as he must have in the late fifties, when the soprano's virgin status disguised the influence of the tenor saxophonist Sonny Rollins on his phrasing and overall sense of improvisational design. Bechet was Lacy's first inspiration. What he responded to, on first hearing Bechet's 1941 recording of Ellington's "The Mooche," was Bechet's *"throb*—the vitality, the intensity, the passion, the sheer sound." In that sense, Lacy, too, has a "throb," although his quicker vibrato and penchant for sustained notes that are implicitly microtonal make it more of a flutter—his sound is centered in his fingers rather than his throat, but is no less seductive or vocal in pitch because of that. He says that Monk once admonished him to "stop playing the piano part"—Monk's typically cryptic way of telling him that his solos were becoming harmonically overcrowded. Lacy obviously took this advice to heart: like Monk, he makes structural use of silence, often seeming to measure rhythm as the distance between notes. Once you've heard Lacy's, there's no mistaking him for anyone else. His timbre and his attack—the way he lands on notes—are brisk, tart, and comically threatening—like a duck calling your name, as the British jazz critic Richard Cook once put it (conscious, no doubt, that Lacy has a piece called "The Duck").

Like that of any musician, Lacy's career can be broken into tidy biographical chapters, but it's the continuity between chapters that gives his work its integrity. Only recently reissued, *The Complete Jaguar Sessions* (Fresco Jazz FJ-1)—a two-record set of Lacy's first recordings, as a member of the trumpeter Dick Sutton's sextet—reveals that there was something suspiciously progressive about Lacy's brand of revivalism in 1954. In contrast to most trad bands of the period, whose players took stubborn pride in how *few* pop songs they knew, Sutton and Lacy's repertoire included "A Foggy Day," "As Long as I Live," and "How about You," to cite just three examples of songs that no New Orleans brass band would think of blowing on its way back from a funeral. Sutton's group had more in common with the original Gerry Mulligan Quartet than with the Preservation Hall Jazz Band: it was pianoless, pivoted around a baritone saxophone, and experimented with shadings and dynamics. Lacy wasn't exactly taking a blind leap when he joined Cecil Taylor two years later to play a nascent form of free jazz, or when he formed a band with the trombonist Roswell Rudd (another former revivalist) in 1962, that eventually played only Monk tunes.

"What is jazz anyway?" Lacy asked rhetorically when I pointed this out. "The first jazz was contrapuntal, ensemble-oriented stuff. So I still play dixieland in a way. The beat, the concept of swing that came out of New Orleans, that temperature, that hot flavor, the spiciness of it all—I try to maintain all of that."

In his two decades in Europe, Lacy has participated in chance improvisations and performed unaccompanied concerts. In the last decade, he has also emerged as an ambitious and prolific composer, which is surprising, given that until he reached forty he was known strictly as a free improviser and an interpreter of Monk, Taylor, Gil Evans, Duke Ellington, and Billy Strayhorn. "I had to find out what music was before I could write my own music," Lacy told me, revealing an analytical bent also present in his improvisations, which tend to examine pitch and intervalic relationships. "I studied Stravinsky, Webern, Prokofiev, Charlie Parker, and especially Monk."

As a composer, Lacy has revived the concept of combining jazz and poetry, although the designation "jazz and poetry," with its bad memories of bongos and berets, hardly does justice to his experiments with music and texts by Lao Tzu, Herman Melville, Samuel Beckett, Blaise Cendrars, William Burroughs, Brion Gysin, Robert Creeley, and Anne Waldman, among others. These have more in common with Steve Reich's settings of poems by William Carlos Williams and Morton Feldman's settings of poems by García Lorca and Frank O'Hara than with Zoot Sims and Al Cohn playing behind Jack Kerouac. Lacy's partner in this endeavor is Irene Aebi, a Swiss-born singer who also plays violin and cello in his sextet and has been his lover since 1966. The task of accommodating a singer has brought about subtle changes in Lacy's music, the most beneficial of which has been a return to something resembling song form. When I mentioned this, he pointed out that he has come full circle as a native New Yorker who was "steeped" in Broadway musicals before discovering jazz. "Words and music fusing to become song. I can do that because I have Irene as a protagonist, an instrument, a voice."

But Lacy hasn't put free improvisation—or his studies of other composers— behind him. Although the pieces he writes for his sextet are no longer openended, there is plenty of room for leeway within them. Lacy still plays solo concerts, and still plays Monk, often unaccompanied. And although it's been almost three decades since he last worked with Taylor, the pianist's music still influences Lacy's own. "His scraps were my vittles," Lacy once told a Canadian interviewer, meaning that he still employs many of the strategies that Taylor has since abandoned (improvisations on an unstated theme, for example, or what Lacy calls "cellular" improvisation on selected harmonic intervals or rhythmic motifs).

Lack of work first drove Lacy from the U.S. in 1966. "The strivers were being driven underground. At the same time, there was a tremendous thirst for freedom, which the music reflected and sometimes instigated. Everyone else was listening to Dylan and the Beatles, and we were listening to them, too. It wasn't jazz, but it was swinging music—swinging maybe more than we were doing. Sometimes the music is going through a difficult stage and only the experts are listening. Everybody else wants to party."

When I spoke with Lacy, he was staying in a suite in a Central Park hotel just a few blocks from where he grew up. After visiting his sick mother, he'd made the rounds of bookstores earlier in the day, and couldn't wait to show the members of his band, who were staying in rooms with windows looking into his across a small courtyard, the Stanislaw Lem books he'd found. He is serious and self-possessed, with a peninsula hairline, Jack Nicholson eyebrows, deep frown lines around his mouth, and a wind player's thick, corded neck.

Before emigrating to Europe, Lacy was stranded in Buenos Aires for a year. He went there with an international quartet, featuring the bassist Johnny Dyani and the drummer Louis T. Moholo, both from South Africa, and the Italian trumpeter Enrico Rava. "Argentina was brutal and oppressive. There had been a series of military coups d'états, and the tanks were still in the streets. Women were forbidden to wear slacks, the Beatles were banned from radio, and we made the mistake of advertising our concert as a 'revolution in jazz.' Our year there was one long tango. We did a lot of rehearsing and a lot of starving, but we did manage to record what I think were the first entirely free improvisations on record" (included on *The Forest and the Zoo*, ESP Disc 1060, long out of print).

Jobs were scarce when Lacy returned to New York in 1967, but he was at least able to perform a radio work that eventually materialized as *The Way*, his six-part suite based on the sayings of Lao Tzu. "It was a pre-version with Irene and Richard Teitelbaum on synthesizer, for WBAI, called *Chinese Food*. In the streets at that time, people were screaming 'L.B.J., L.B.J., how many babies have you killed today?' We wanted to scream at him, too, but we wanted to do it musically. So we did a sort of improvised cantata, with Irene shouting the more political parts of the Tao. That was the genesis of *The Way*, which was my way into composition." (The most fully realized version of the work is the 1979 Swiss concert performance on hat HUT THREE [2R03].)

Lacy and Aebi next went to Rome, where for three years they performed "with groups mostly made up of amateurs, because this was a period in which audiences were being invited up on stage with actors in an effort to break down the fourth wall, and we were into that, too. But that becomes frustrating after awhile. I was writing music like I'm writing now, but I couldn't find anyone to play it. We went to Paris because I knew there would be good musicians there."

As a musician with a keen interest in developments in the other arts, Lacy's experience as an American in Paris has been more like that traditionally associated with novelists and painters than that of his fellow jazz expatriates. "It was that other stuff that attracted me, too—all the arts with a capital 'A'," he told me. "I like to rub up against it." The audience for Lacy's music in Paris isn't much larger than it was in New York, but, as Lacy pointed out—drawing an analogy to painting—"You're considered a failure if you have an empty gallery in New York. In Paris, you're not." (This is a variation on a line that Keith Carradine delivers to explain the appeal of Paris to artists in Alan Rudolph's *The Moderns*: "It's okay to be broke in Paris. In America, it's downright immoral.") Lacy's larger works, which have involved dance, visuals, and texts (works such as *Futurities*, *Brackets*, and *Naked Lunch*) might not have been realized in the U.S., where jazz remains somewhat isolated from the "fine" arts by virtue of its low pedigree and ancestral ties to show business. (Lacy could be called a "French" composer. Earlier this year, he was commissioned to write a piece for the French bicentennial.) Because Lacy is white, Europe isn't a haven from racism for him, as it has been for so many black American musicians (and because he's white, he didn't have a ready-made audience of French jazz fans for whom blackness is the only proof of authenticity). Still, though Lacy says otherwise, I would suppose that race played at least a small part in his decision to leave the States at a time when free jazz was being equated with Black Power and white musicians were unwelcome on some bandstands. And something he said when I asked him what it would take to bring him back home suggests that race plays a part in his decision to remain in France. "You'd have to uproot my band, and my drummer and piano player might not want to come," he said. I doubt it was coincidence that he referred to two of the three black expatriates among his sidemen.

Keeping up with new developments back home and finding competent European sidemen—these are the challenges facing the typical jazz exile. But the first is irrelevant to Lacy, who has never really been in the mainstream anyway, and the second has long since ceased to be a worry: he has kept a band together since the early seventies, with no changes in personnel since the expatriate American pianist Bobby Few joined in 1981 (the group's other members, in addition to Lacy and Aebi, are the alto and soprano saxophonist Steve Potts and the drummer Oliver Johnson, both also expatriates, and the French bassist Jean-Jacques Avenel). Paris is Lacy's gateway to the rest of the Continent. He performs with his sextet throughout Europe, and returns to the U.S. for a three-or-four week string of engagements once or twice a year. He wouldn't necessarily be in greater demand in the States if he lived here year-round, and he would probably have to spend a good deal of time in Europe anyway, because that's where most of the work is for jazz musicians now.

Europe is also where the recording opportunities are, as Lacy's bulging European discography confirms: he averages a half dozen or so albums a year for foreign labels. In the last two years, these have included (in addition to Lacy's dates as a sideman) the solo albums *Hocus Pocus* (Le Disques du crépuscule TWI-683), *Only Monk* (Soul Note SN-1160) and *The Kiss* (Lunatic 002); *The Condor* (Soul Note SN-1135) and *The Gleam* (Silkheart SHLP-102) by his sextet; *The Window* (Soul Note 121-185-1), with just Jean-Jacques Avenel and Oliver Johnson, the sextet's bassist and drummer; *One Fell Swoop* (Silkheart SHLP-104), with Avenel and Jackson, plus the alto and baritone saxophonist Charles Tyler; three solo albums (including one of Monk tunes); free improvisations with fellow soprano saxophonist Evan Parker (*Chiros*, FMP SAJ-53) and duets with Potts (*Live in Budapest*, Westwind 011) and with the pianists Gil Evans (*Paris Blues*, Owl 049), Ulrich Gumpert (*Deadline*, Sound Aspects SAS-013), and Mal Waldron (*Sempre Amore*, Soul Note SN-1170). But serendipitously, Lacy's strongest recent releases are those that will be easiest to find in U.S. record shops.

Momentum (RCA Novus 3021-1-N) and *The Door* (RCA Novus 3049-1-N) are his first American releases since 1977, and his first ever for a major U.S. label. Both feature his sextet and demonstrate his success in forming a band in his own likeness. Although Steve Potts's scattershot soprano solos are the antithesis of Lacy's, his alto solos sound like fatter and juicier versions of Lacy's soprano ones. Bobby Few is a splashy pianist given to heavy chording and semiclassical allusions; he should be all wrong for Lacy's band, but his density and extravagance effectively counterbalance Lacy's austerity. Avenel and Johnson are an impeccable, close-knit rhythm team.

But besides Lacy himself, the member of his group who best defines its sound is Aebi, whose singing—a not-always-blissful marriage of *sprechstimme* and scat—takes some getting used to. In the past, Aebi has often sounded uncomfortable singing in English. On *Futurities* (hat ART 2022—the only one of Lacy's large-scale collaborative works available on record), Aebi's absurdly trochaic declamation undermined the offbeat colloquialism of Robert Creeley's poetry—though, to be fair about it, Lacy was also to blame for the rigidity of his settings. (The bassist Steve Swallow's 1979 attempt to set Creeley to music was similarly unsuccessful due to fixed meters. What I would like to hear is a reading that Creeley gave for French radio with Lacy improvising behind him.) Aebi's singing also forces her to put down her strings, thus depriving the group of the illusion of a third horn that she creates with her deliciously sour violin.

Momentum's vocal tracks—Brion Gysin's translation of an Islamic poem, Herman Melville's "Art," and Giulia Nicolai's "Utah"—are surprisingly gratifying, perhaps because the texts are less familiar and one has no preconceived notion of how they should be declaimed, or perhaps because Lacy is becoming steadily more confident in exploiting Aebi's theatricality. "Utah" is delightful: a singsong litany of colorful Utah place names set to one of Lacy's jauntiest melodies. (This is one of two tracks included only on compact disc. The other is "The Gaze," a requiem for Marvin Gaye, with grimly chuckling horns and a title that suggests Lacy has been reading the French semiologists.) The most finely wrought of the instrumental tracks is "The Bath," a blues in which Lacy and Potts converge in slightly sharp unison on the main theme, before pulling apart on the canonlike bridge. "The Bath" exemplifies another valuable lesson that Lacy absorbed from Monk: "He used to tell me the inside of a tune was what made the outside sound good. I've never heard a better explanation of the function served by writing a bridge."

The Door, which is all instrumental, includes themes by Monk, Ellington and Strayhorn, and George Handy, in addition to three compositions of his own that Lacy has recorded before in different settings. The full sextet performs only on Ellington's and Strayhorn's "Virgin Beauty," on which the late Sam Woodyard, Ellington's drummer from the mid-fifties to the late sixties, also makes a cameo appearance, combining with Jackson for an irresistibly exotic groove. Avenel provides an effortless, hammocklike swing on Monk's "Lazy Beauty"; Potts deftly expands a simple rhythmic motif into austere complexity on Lacy's "Blinks"; and Few sparkles throughout the album. But Lacy himself is the reason this is a memorable album. He does his best work on two duets. On his own "Clichés," as Avenel creates an appropriately hypnotic background on sanza (African thumb piano), Lacy quietly plays split and overblown notes most other saxophonists would find it necessary to scream. The track is a veritable inventory of the freak effects the saxophone is capable of producing in the aftermath of free jazz, but it's also an exercise in subtlety. So is Lacy and Few's interpretation of Handy's "Forgetful," a lovely ballad first recorded by the Boyd Raeburn Orchestra in 1945. Considered experimental for its time but somewhat dated now, the Raeburn recording surrounded David Allyn's Sinatra-like vocal with cackling dissonance that Lacy wisely ignores in penetrating to the song's melodic heart. Avoiding sentimentality by calling attention to the song's steep melodic progressions and by exploiting the soprano's natural tendency to buzz, Lacy offers little more than straightforward embellishments. But in terms of the vitality and character it projects, this is a modern ballad interpretation to place alongside those of such old masters as Johnny Hodges and Ben Webster.

It's one of Lacy's finest recorded performances, in what has been an exemplary career.

(*November, 1989*)

THE PHILADELPHIA STORY

Is there a jazz sound unique to Philadelphia? In lieu of a style or movement indigenous to the city (as Dixie was to New Orleans, stride to Harlem, and swing to Kansas City), the civic boosters who say yes offer a litany of the famous musicians born or bred in Philly. It's up to the rest of us to point out that it would be remarkable if a city of Philadelphia's size and racial makeup *failed* to produce an abundance of jazz talent. So perhaps the question needs to be rephrased. Are there any characteristics historically identified with Philadelphia-area musicians? The answer to that one is yes, and the tenor saxophonist Odean Pope—familiar to listeners around the globe from his work with the drummer Max Roach's quartet and from his own releases on Moers Music and Soul Note—is a perfect illustration.

Like so many Philadelphia musicians, including John Coltrane, Pope arrived in the city shortly after World War II as part of a great black migration from the Carolinas. As is the practice in most cities, he apprenticed with older and more experienced musicians (some of them unknown outside city limits), and played rhythm 'n' blues (behind James Brown, Marvin Gaye, Stevie Wonder and countless others, as a member of Sam Reed's house band at the Uptown Theater, Philadelphia's equivalent of the Apollo). What really distinguishes Pope as a Philadelphian, however, is his quasi-spiritual obsession with harmonic theory and instrumental technique. This is what he has in common with Coltrane, the vibraphonist Walt Dickerson, and the guitarist Pat Martino, among others. For such musicians, the instrument becomes an extension of the body; music, a manifestation of the soul. Becoming a better musician therefore means becoming a better human being, giving new meaning to the old saw, "practice makes perfect." "I think it goes back to when Dizzy Gillespie was still here, in the late thirties and early forties," Pope says. "Dizzy's technique and harmonic imagination were so advanced that the local musicians who came after him felt like they had to spend a lot of time in the woodshed in order to maintain his standard."

Pope, whose father was a semipro baseball player, and whose mother was a Sunday-school teacher and church choir director, was born in Ninety-Six, South Carolina, in 1938. "It was a little village with two major streets, the railroad tracks on one side and stores on the other. You could walk through town in ten or fifteen minutes. "Pope, who sang in his mother's choir as a child, feels that his groundswell rhythms and oratorical inflections derive—much as Coltrane's were said to—from early exposure to Southern Baptist preachers: what he calls "the spiritual thing" (a better name for it than "the Southern thing," because it became a Philadelphia characteristic as generations of local-born musicians exchanged ideas with the newer arrivals from the south).

Although Pope studied at the Granoff School of Music and names Phila-delphia tenor saxophonists Bill Barron, Wilber Campbell, and Jimmy Oliver as early influences, his most important tutor was Hasaan Ibn Ali—*The Legendary Hasaan*, as the late pianist was billed on his one album. Hasaan was Phila-delphia's answer to Thelonious Monk, a maverick who seemed a bit odd even to musicians accustomed to eccentricity. "Hasaan was so advanced that musicians shied away from him. Max Roach was the only one who ever gave him a break. [Roach sponsored and played drums on the trio session that Hasaan recorded for Atlantic in 1965; another session with Roach and Pope from the same year was never released.] Hasaan never had a day gig in his life," Pope remembers. "He would practice all day in his bathrobe, from the morning on. His father would bring him breakfast, lunch, and dinner at the piano. At night, after he got dressed, there were three or four houses he would visit, where they had pianos. The people would serve him coffee or cake, give him a few cigarettes or maybe a couple of dollars from time to time. He was very dedicated, very sincere, but also very outspoken, and other musicians some-times didn't know how to deal with him. If someone wasn't playing well, he'd come right out and say 'Man, I think you need to study some more.' He had a bad reputation among pianists. If we were in a club, and the pianist wasn't making it, Hasaan would push him right off the bench and start playing himself. It got so that when a piano player saw him coming through the door, he got up because he knew he was going to be getting up anyway.

"He was one of the outcasts in Philadelphia, and for a while, musicians ostracized me, too, because of my association with him. Nobody ever gives Hasaan any credit, but every important musician who came out of this area in the fifties and sixties, including McCoy Tyner, learned from him. He's the one who showed McCoy the G-minor seventh chord, which is so popular today, thanks to McCoy. Hasaan knew so many ways of approaching chord changes, so many different substitutions. For example, the C-seventh chord: he would approach it from the seventh degree, the fifth degree, the flat-ninth. The triangle major-seventh, which Coltrane used on 'Naima' and that everybody uses now—I think it was Hasaan who showed that chord to Trane."

Hasaan showed Pope quite a bit, too, during the period from 1960 to 1965, when Pope "stopped accepting even the few gigs I was being offered in order to study and concentrate on what I was hearing in my head." Learning that Pope was also studying the oboe, Hasaan suggested that he might try using oboe fingering techniques on his tenor saxophone. Pope still employs oboe "fork fingering" to achieve difficult overtones and multiphonics, "three, four, as many as five notes at a time," on the tenor saxophone, an instrument long presumed incapable of producing chords. Because Hasaan emphasized the importance of originality, Pope discarded all the albums he'd acquired by tenor saxophonists, for fear of undue influence. He listened instead to "all the great pianists—Hasaan, Monk, Fats Waller, Earl Hines, Bill Evans." Out of admiration for their long, unbroken keyboard lines, he developed a system of circular breathing that allowed him to play as many as three choruses without coming up for air.

Pope stills likes to practice "at least three or four hours a day," usually in a studio at the Settlement Music School, where he taught saxophone and improvisation and directed the student orchestra from 1976 to 1979. "It's quiet—no telephone, nothing to interrupt me. I concentrate on reading, timbre, intonation, harmonic theory, and improvisation using scales: five-tone scales, seven-tone scales, nine-tone scales, eleven-tone scales, the scales that enable you to play advanced harmony when you improvise. I get cranky when I'm not able to practice," he says. Although he twice tried living in New York as a younger musician, he refuses to move there now for fear that "the hustle of paying New York rent and New York prices" would distract him from his study. In 1979, his dedication to musical self-improvement almost caused him to pass up an offer to rejoin Max Roach, the great drummer with whom he'd worked briefly a decade earlier. "Between teaching at Settlement four days a week and playing with Catalyst [a Philadelphia-based cooperative band, now defunct except for periodic reunions], I figured I was already too busy. I wanted to be sure I had time to study and compose. Fortunately, I have a sensible wife, who said, 'Look, Odean, to play with a musician of Max's stature is an golden opportunity that might not come again.'"

Since joining Roach, Pope has led two lives, playing first-class venues with the drummer, but scuffling for work in dives at home. As the vibraphonist Khan Jamal, another internationally visible Philadelphia musician caught up in the same contradiction as Pope, once observed: "There are plenty of clubs around town I can work in, if I choose. But clubs don't want anything too deep. If I want to play creative music, I have to hustle to line something up at the Painted Bride [a local alternative arts space] or one of the colleges. You almost have to apply for a grant every time you want to play." Around Philadelphia, Pope plays with the Magnificent Seven (a local all-star band) and leads several groups of his own, including a near-harmolodic trio with the bassist Gerald Veasley and the drummer Cornell Rochester. But Pope's most ambitious outlet is the eleven-member Saxophone Choir, which makes its debut on *The Saxophone Shop* Soul SN-1129).

"I guess it goes back to the church music, the desire for massed voices, and the fact that saxophones are the instruments closest to the human voice," Pope says to explain the Saxophone Choir's unusual instrumentation—eight saxes plus rhythm. The glory of *The Saxophone Shop* is that it sounds like Odean Pope multiplied to the eighth power. As an orchestrator, he's elongated the already lengthy lines he favors as an improviser without stretching them too thin. The elegiac "Cis," reprised from *Easy Winners* by Roach's Double Quartet (the regular unit plus strings), is the album's outstanding cut, but the pieces taken at brisker tempos convey breathtaking motion and sweep. "Almost Like Me—Part II" borrows both its name and its jaunty gait from Hasaan ibn Ali, who also lurks behind the compound rhythms of "Mantu Chant" (let's remember that Hasaan once titled a piece of "Three-Four vs. Six-Eight Four-Four Ways"). Although Pope is the only soloist given much blowing room, several other Philadelphia musicians shine in supporting roles. The pianist Eddie Green cuts through the horns with intelligence and style. The drummer Dave Gibson nips at Pope's heels on "Prince La Sha," and the tenor saxophonist Bob Howell duels him to a draw on "Almost Like Me—Part II." The alto saxophonist Julian Pressley's elegant choruses buoy Pope's arrangement of Clifford Jordan's "Doug's Prelude," and on "Elixir," Veasley's skillful, flamencolike strumming sets the stage for an unaccompanied Pope solo that is an example of Philadelphia jazz at its most spiritual and harmonically engorged.

(*April, 1987*)

STREAMS OF
CONSCIOUSNESS

Sporadically throughout his life, the fifty-one-year-old pianist Ran Blake has kept a bedside diary of his nightmares—"usually only for six months or so at a time," he says, "until I begin to feel ashamed of my self-indulgence. Besides, I'm so in touch with that state, so able to recall the imagery during waking hours, that keeping a journal is a bit superfluous. Sometimes, following a particularly disturbing dream, I rush to the piano to recapture the mood in composition; the notes are already there in my subconscious. I had a real lulu the other night. I was at my own funeral, but I wasn't in my grave. I was there as an observer, watching the mourners interact with a group of strangers enjoying themselves around a bowl of punch that someone had brought out to the cemetery. I became very caught in this, almost the way I do in certain films."

Blake's dreams are in black and white, like the keyboard at which he labors, like the racially checkerboard jazz subculture he inhabits for lack of a more suitable niche, like the *noir* films (those of Alfred Hitchcock, Fritz Lang, and Robert Siodmak, in particular) that rescued him from loneliness as a child in New England in the 1940s. "As a teenager, I spent a few hours a day practicing the obligatory scales, hating every second. But I would sneak off to the movies two or three times a week, and then creep down to the living room piano in the dead of night—careful not to play all the dynamics, so as not to wake my parents—and attempt to convey my impressions of the films while the memory of them was still vivid.

"I started collecting soundtracks, but soon realized that it wasn't film music that gripped my imagination—it was the films themselves, and their ambiance more than their plots: Dana Andrews slowly falling in love with the painting of Gene Tierney in *Laura*, for example. Except for some scores by Bernard Herrmann, the music generally wasn't rich enough for me, unless there was the implication of violence or foul play, or unless the characters were experiencing sensations of fear, guilt, anxiety, or dread, which the music had to establish. Then there might be a few dissonant chords that appealed to me, but nothing that Bartók and Stravinsky hadn't already done better. If I had only known about them, I could have been studying twentieth-century composers instead. But I spent every spare moment as an adolescent in movie houses and black churches.

"One Sunday morning when we were still living in Springfield [Massachusetts], my parents sent me off to services, and I took a wrong turn and wound up at a black Pentecostal church, beckoned there by those pounding rhythms. I went back every Sunday after that. When my folks asked me how church was, I'd say I loved it, and not be lying—exactly. This was my first exposure to black music, and it lasted for months—until my parents ran into our pastor. When we moved to Suffield [Connecticut], which was lily white at the time, I would travel to the Holy Trinity Church of God in Christ, in Hartford, where I wound up making my professional debut, playing for the gospel choir. I remember they said they like my rhythmic feel, though they had some qualms about my dissonant voicings.

"So there you have it: *film noir*, Mahalia Jackson's moan, and the midnight world of dreams. Those have been the chief influences on my music."

Despite winning the approval of the congregation in Hartford—to say nothing of the skeptical audience at Harlem's Apollo Theater, where, in 1961, he and the black singer Jeanne Lee won an amateur-night competition with their nubby, decelerated interpretations of such pop standards as "Laura" and "Summertime"—Blake felt out of place in black jazz circles. "White jazz was Stan Kenton and Gerry Mulligan; black jazz was Thelonious Monk, Charles Mingus, Max Roach, and Abbey Lincoln. You can guess which I gravitated to," says Blake. Shortly after arriving in New York, he took a job as a waiter at the Jazz Gallery, but was demoted to the kitchen after he "became so involved in what was going on in the bandstand that I fell over people's legs and dropped a tray of drinks on James Baldwin's lap. The Baroness Nica de Koenigswarter saved my job, and I was eventually made her private waiter. The music I wanted to play had black roots, but I was approaching it from a white-intellectual perspective, which put me at too great a distance from the source. All the same, I had no desire to become one of the hip young white boys sitting in at Birdland every Monday night. I wasn't interested in blowing twenty choruses on the chord changes of 'All the Things You Are,' even if I could have—and, believe me, I couldn't, because I would grow bored, start to daydream, and miss the turnarounds. I didn't much like playing with bassists and drummers, and they absolutely dreaded playing with me. I didn't read well enough to become a classical pianist, and much as I loved singers, my chord choices were all wrong for vocal accompaniment. I was forever running to Bill Evans, Oscar Peterson, and Mal Waldron for lessons and career counseling, and at one point, in the late sixties, I almost called it quits. I never stopped playing, but, from 1971 to '76, I stopped hustling for gigs, though I did take what few gigs were offered me. Oddly enough, the few musicians who thought I had something new and provocative to offer were black, but their approval never reached the point of hiring me for their bands. They knew that what I was playing wasn't compatible with what they were doing."

Blake credits the composer, conductor, critic, and educator Gunther Schuller with "saving my life by suggesting that there was more than one way to approach improvisation. Maybe my music wasn't jazz at all, he said. Maybe it was 'Third Stream,'"—the phrase Schuller had coined in the fifties to describe the confluence of jazz and classical musics. Schuller, appointed dean of the New England Conservatory of Music in Boston in 1967, named Blake to the extension faculty a year later. Initially Blake was in charge of the school's community outreach program (in the inner city, prisons, senior citizens' homes), but in 1973, Schuller created a degree program around him. As chairman of the Department of Third Stream Studies, Blake has gradually broadened Schuller's original definition of "Third Stream" to include temporary alliances of Western and ethnic musics—"sometimes bypassing jazz and classical altogether," Blake notes. Indeed, the best-known group to emerge from Blake's classroom is the Klezmer Conservatory Band, which usually plays Jewish community centers and synagogues rather than concert halls or nightclubs. (Hancus Netsky, the band's director, is now chairman of the New England Conservatory's jazz department.) "There's streaming in Mingus— hard jazz, lush balladry, church music. But he'd probably have punched you in the nose if you were to call his music Third Stream," says Blake, who teaches "by ear, as in the African aural tradition," and whose students are known around Boston as "Also-Rans."

"Students in the Third Stream Department learn to create a highly individual music," reads the Conservatory's course guide, "a music they feel in themselves but do not hear around them"—which is precisely what Blake has done in his own music, in making Third Stream a stream of consciousness. The records of his that count are those he's made since turning forty (although it would be a mistake to dismiss his earlier work out of hand, as an Italian critic recently did, in a burst of enthusiasm for his output since 1976). These later records include *Duke Dreams* (Soul Note SN-1027: dark reflections on the corpus of Ellington and Strayhorn), *Suffield Gothic* (Soul SN-1077: a meditation on New England as repository of personal memory and national myth), and *Film Noir* (Arista Novus AN-3019) and *Vertigo* (Owl 041)—companion attempts to retrieve the frisson of cinematic melodrama from the flicker of memory).

Musically, Blake is content to travel his own path and let the world catch up with him when and if it chooses. But he seems beset by insecurities of a more personal nature. When I spoke with him before a concert in Philadelphia in April, he told me that several thousand had turned out to hear him in Greece a few months earlier. "But I doubt it was me who drew them," he said, scrunching his face. "They were just curious to see an American musician." In Philadelphia, the crowd was in the dozens, and Blake feared that the promoters were losing money by indulging their personal enthusiasm for him. Told that Sun Ra and Cecil Taylor were scheduled to play in the same solo piano series, he was more alarmed than flattered: "I feel a bit like an imposter, because their music is much closer to jazz than mine is." And although he is a tireless recruiter for the New England Conservatory (the Third Stream Department is allowed to continue only so long as enrollment demonstrates a need for it), he voiced doubts about the value of jazz education, pointing out that "Monk and Bix Beiderbecke did pretty well without it."

I asked Blake if he still considers himself a political artist (in 1969, he released *The Blue Potato and Other Outrages* . . . [Milestone M-9021], an album of dedications to Eldridge Cleaver, Malcolm X, Che Guevera, and Régis Debray, combined with standards whose titles commented ironically on Southern racism and the military coup in Greece). "No," he said. "I may be an impressionist, but my music is not programmatic. Audiences have no way of knowing if a thundering cluster is supposed to represent police brutality in South Africa or the dishes falling off my table as I attempted to prepare a curry the night before. Maybe if I were famous, it would be different. Nobody much cares what I think, and it's too easy for a white man to exploit black issues for his own self-aggrandizement. I guess I feel more impotent against injustice than I used to. At a certain point, I realized that listening to Billie Holiday, enjoying a reasonably good dinner, and reading a stimulating book an hour before bedtime to facilitate entry into the dream world are the activities that have the most bearing on my music. I know that sounds selfish, but I don't want to pass myself off as better than I really am."

For his Philadelphia concert, Blake played Monk, Fletcher Henderson, Shorty Rogers, Pete Rugolo, John Philip Sousa, Bernard Herrmann's score for *Vertigo* enfolded with Blake's own variations, adaptations of traditional Sephardic music inspired by the recent film *Shoah*, and impressions of *Rebecca*, *The Wild One*, and *The Wrong Man*—a characteristic program for a performer who has raised eclecticism to a discipline. The surprises were Rogers and Rugolo, palefaces associated with West Coast jazz and the top-heavy Stan Kenton Orchestra of the forties and fifties (but, significantly, the most obvious precursors to the Third Steam movement, aside from Ellington and Mingus).

Blake will teach a course on Monk this fall, which should prove interesting, because Monk is the jazz pianist he most resembles, if only in the emphasis he places on the nuances of *touch*, an aspect of pianistics criminally overlooked in most assessments of technique. "It's funny. I once wrote an article on Thelonious for one of the keyboard magazines, and never even mentioned his touch, though I certainly should have," says Blake, who was once the jazz critic for *The Bay State Banner*, a black-owned newspaper in Maine. "It's also something I've never given much thought to in my own playing, though I do remember one of my first teachers telling me not to bang the keyboard, and me thinking that there were instances when banging was called for. I play fewer notes than most pianists, and perhaps that's Thelonious's influence. But I think it has more to do with a subconscious desire to imitate a vocal line. I'm probably the only pianist of my generation more in debt to Billie Holiday, Mahalia Jackson, Abbey Lincoln, Chris Connor, Stevie Wonder, and Victoria de los Angeles than to Bud Powell."

(August 1986)

POSTSCRIPT

Tour de force, with its implication of facile ingenuity, isn't a phrase I would normally dream of using in connection with Ran Blake: a musician who, to his credit, makes a point of emphasizing substance over style. Nonetheless, "Short Life of Barbara Monk," the title track of Blake's new album (Soul Note SN-1127) strikes me as a *tour de force* on every level. With a main strain appropriated from a children's song (and swung to a $\frac{6}{4}$ rhythm that sounds like a waltz veering out of control), the piece is dedicated to the memory of Thelonious Monk's daughter, a member of the funk group T. S. Monk. Her death (from cancer) followed that of her father by less than two years. Like many of Blake's best compositions, it owes its existence, at least in part, to a dream. "I see her ice skating on a winter day in New York as a little girl," says Blake, who once was a sort of baby-sitter for Barbara and her brother, Thelonious, Jr., in the aftermath of a fire that gutted Monk's home. "Though baby-sitting is probably too glorified a word for what I did. I offered to help in any way I could, and I'm sure I was more of a nuisance than a help. I wasn't even sure how you changed a light bulb, and I think Thelonious got a kick out of me. He was a very private man, and I was too in awe of him to engage him in the sort of musical conversation he must have had with disciples like Barry Harris.

"But believe it or not, I never did see her ice skate. It was just a dream I had after hearing that she had died." In its programmatic juxtaposition of timeless folk gaiety and modern, brittle atonality, "Short Life of Barbara Monk" resembles certain of George Russell's early works, most notably "All about Rosie" and "The Day John Brown Was Hanged." (Although the resemblance may be unintentional, it's not surprising in view of Blake's longstanding admiration for his fellow NEC faculty member.) But with its contradiction of moods (to say nothing of its Bernard Herrmann–like dramatic pull), the piece also seems to be yet another refection of Blake's fascination with the way music works in *film noir*, as a clue to ambivalences and anxieties that can't be expressed in dialogue, *mise en scène* or actors' bits of business.

"Short Life of Barbara Monk" is an incredibly rich work, and not the least of its virtues is Blake's disjunct but thematically apposite piano choruses over pulsating bass and drums. Which raises a point I doubt will be lost on reviewers: a Ran Blake album that begins with a walking bass is something of a shock. There are two possible explanations why Blake waited this long—twenty-five years into his recording career—to make a quartet date with saxophone, bass, and drums. One is that genius frequently overlooks the obvious. The other is that Blake's rhythmic insecurities long precluded such a simple presentation. "I wanted to make sure that I would play with musicians who heard notes before their fingers touched their instruments, and these musicians do," he says of bassist Ed Felson and drummer Jon Hazilla, both Conservatory graduates. The quartet's fourth member is tenor saxophonist Ricky Ford, another of Blake's former students, whose free but carefully focused solo on *Short Life of Barbara Monk* rivals his work with Abdullah Ibrahim as his most satisfying on record. Because Ford, Felson, and Hazilla are so attuned to Blake, what might have been a retreat into the conventional instead becomes a vindication of Blake's peculiarity.

Although the title track fully deserves its place of honor, the other performances here are just as outstanding in their various ways, and they provide insight into the wide scope of Blake's current enthusiasms and passions. Cole Porter's "I've Got You Under My Skin," which Blake associates with the singer Chris Connor, and the two takes of the traditional Sephardic theme "Una Matica de Ruda," which he first encountered on a record by Victoria de los Angeles, testify to his love for vocal music. Two items redeem material from the Stan Kenton repertoire: Bill Russo's mambolike "23 Degrees North—82 Degrees West" and the Kenton Orchestra's signature theme, the bandleader's own "Artistry in Rhythm." In both instances, Blake excavates the metric vitality buried under all that brass (and Ford lets loose with an especially forceful solo on the Russo piece). Helped in no small measure by Ford's ability to evoke Ben Webster and Paul Gonsalves without mimicry, "In Between" and "Dark" written by former Blake students Claire Ritter and Mauricio Villavecchia, respectively) celebrate the impressionism introduced to jazz by the Duke Ellington Orchestra through the early works of Billy Strayhorn. "Vradiazi," though retaining its mood of turmoil, is the most buoyant of the six versions of this Mikis Theodorakis piece that Blake has recorded. Finally, there are two Blake originals related to the title track if only in inspiration: "Impresario of Death," a seven-note theme suggested by an oppressive, disquieting dream; and "Pourquoi, Laurent?" a stricken response to the suicide of the French jazz critic Laurent Goddet. Inasmuch as death (including the recent death of his mother) is the "subject matter" of this remarkable album, it's a tribute to Blake's complexity as a composer and performer that the tone is so doggedly life-affirming—in the manner of all great music.

(November, 1987)

In 1988, Blake was awarded a MacArthur Foundation "genius" grant for $320,000 over five years.

CIRCLES, WHIRLS, AND EIGHTS

"I had this incredible vision when I was younger," the pianist Borah Bergman said recently. "As I'm playing, everyone is running out of the hall, and this is making me very, very happy. I had a friend around that time who sometimes played drums with me, and I once asked him, 'What do you think I do best?' He looked at me and said, 'Borah, the thing you do best is make tumult.'"

Bergman is an outcat. The hipster pianist Paul Knopf, who released three excellent but overlooked albums in the late fifties, coined the word, in a mood of flippant defiance, to celebrate himself as "an outcast and a far-out cat combined." By popular stereotype, all jazz musicians are outcats. But those of us within music recognize the outcat as a specific type too self-absorbed to be part of any movement (and too idiosyncratic to spearhead one, even post-humously), and too self-reliant to seek audience or peer approval (and too marginal in the larger scheme of things to elicit much). The word conveys undertones of exile, rootlessness, alienation, despair. Thelonious Monk? Maybe. Herbie Nichols? Definitely.

Outcats tend to be pianists, which seems only fitting, because the keyboard—a complete orchestra within hand's reach, as most outcats, including Bergman, think of it—is just the instrument for a musician determined to keep his own counsel. Though they seldom play jazz, there are outcats in movies (Frank Sinatra in *Young at Heart* and Harvey Keitel in *Fingers*) and in literature (Eddie Leen, the doomy, on-the-lam-from-life hero of David Goodis's *Down Here*, a.k.a. *Shoot the Piano Player*—and Earl Morgan, the father of Virgil Morgan, the narrator of Scott Spencer's *Preservation Hall*).

Spencer's description of Earl Morgan's "lurching, pouncing, halting, racing experiments in pure sound," and Virgil Morgan's alienated response to them, suggest the similar trials that Bergman can put even an experienced listener through when he leans bodily into the piano, with his left hand moving furiously, independent of his right and ultimately usurping, in devil's crossed-hands motion, the right hand's domain at the top of the keyboard: "There was something grand and assertive about (his) compositions and you felt at once lonely and besieged when you heard them. They were pieces to take the color out of stained glass, something to remind you that there is no afterlife."

"A girl who was up here once told me 'Borah, your music isn't *nice*.' But I don't think you have to *like* art. You have to respond to it," says Bergman, fifty-two, who makes his living teaching music in the New York public school system. Originally from Brooklyn, he now lives alone near Central Park, in a studio apartment at once spartan in furnishing and cluttered because of the piano in the middle of the floor and the clavier (a soundless keyboard good for improving finger dexterity) beneath the one row of windows. During a three-hour conversation, Bergman (who looks something like Dick Miller, the bristly comic lead in Roger Corman's *Bucket of Blood*) ran his fingers through his thick black hair so often that it stood up above his forehead in a rooster's comb. He speaks in a hectoring whine, with a crescendo somewhere in each sentence—a speech pattern that shrieks "Brooklyn" to New Yorkers, but shouts "New York" to everybody else. He talks about himself incessantly, revealing insecurity rather than a surfeit of ego; you get the impression that at any moment he might say, 'But enough about me. What don't *you* like about my music?'"

When talking about music, Bergman's thoughts keep veering off to subjects that at first seem irrelevant, until one realizes that the novelists he read as an English major at New York University ("I was always drawn to the expressionistic writers like Thomas Wolfe—it's like Henry Miller said, the important thing is to get it all out") and the abstract expressionist school of painters he admires ("someone once accused me of putting Fats Waller and Cecil Taylor into a shredder and coming out with a solid block of sound—which is a perfect description of a Jackson Pollack painting, by the way") have shaped his music as much as other musicians have. As musical influences, he cites the boogie-woogie and stride pianists, Thelonious Monk, Ornette Coleman, Bud Powell ("I loved his purposeful fumbles"), and Lennie Tristano ("the father of free piano, but he drew back; he never learned to play those lines he created by overdubbing in real time").

Bergman played clarinet as a child, and didn't start piano seriously until he was in his twenties, "when I already knew a good deal about jazz, which was both good and bad. I knew there was no point in sounding *almost* as good as Bud Powell. I said if you're going to play piano, you have to get something of your own; otherwise, quit." He decided to play as fast and hard as he could, for as long as he could. His inspiration was John Coltrane's "Chasin' the Trane": "There may have been sexual innuendoes to it; wanting, unconsciously, to have intercourse that lasted twenty minutes. I don't know. But if you want to think of it in those terms, most pianists, including Cecil Taylor, were limiting themselves to foreplay. . . .

"I felt something going on when I was playing fast. But sometimes, after awhile, I felt nothing. It was all worked out in my mind so I knew what was coming next, and I was *bored*. But I found that if I broke it up, and shot both hands this way and that way, it would revive me." He resolved to learn to play everything he could play with his right hand with the left. Eventually, he began to execute improvised passages and conceive entire pieces for the left hand alone.

Although numerous classical composers, including Ravel and Scriabin, have written pieces for the left hand, the strength and independence of Bergman's left hand makes him unique among improvising pianists. If his music resembles that of any of his contemporaries, it is Cecil Taylor's but only in terms of sound mass and his fondness for recurring motifs. "When the left hand is playing one thing and the right hand another, there are two things going on simultaneously," Bergman says. "So I don't see the comparison to the one very beautiful thing going on in Cecil's parallel-hands playing. I'm right-handed, but I intentionally developed my left hand because I get a completely different feeling in my body from playing with my left hand. I don't know if it's neurological or the nature of the piano, where the left hand is always playing in the darker-sounding areas. There are also cultural implications for me. The left hand has traditionally been the bastard, the stooge, the devil, but also the dreamer. It sounds corny, but I associated the left hand with the oppressed. I'm from a very left-wing background. My parents, who were Russian Jewish immigrants, were socialists, and we were surrounded by anarchists, Trotskyites, and Zionists. I grew up thinking that there was nothing like having a good fight for your rights on your hands. Something to fight for was something to live for. I've always been inclined to go the opposite way. If somebody else had been doing what I was doing with my left hand, I wouldn't have done it. For me, the left hand was the great equalizer. It didn't matter anymore whether I was young or old, or black or white, because I had something of my own."

In choosing such an odd and physically strenuous style, Bergman has sometimes been guilty—much like Norman Mailer, another of his favorite writers—of bidding the will to do the work of the imagination (will is sometimes also required in reading Mailer and in listening to Bergman). But beginning with "Poignant Dreams" and "Ballad of a Child," two suspended-animation ballads on his 1985 album, *Upside down Visions* (Soul Note SN-1080), his work has undergone a thaw similar to the one that Virgil Morgan notices in his father's music toward the end of *Preservation Hall:* "It was different from the kind of music I associated with him. The jittery rhythms had been smoothed out and though the chords were stuffed with notes they seemed remarkably agreeable, almost placid, unlike his usual chords in which the notes blindly warred . . . the music sounded natural, strong, and, as lovely melodies always sound to people who don't understand music, unbearably sad. I had always thought of the great composers brushing teardrops from the score as they worked and now, for the first time, I thought of (him) composing the same way."

There is more breathing room between phrases in *Upside Down Vision's* ballads (and in the title track, taken at an only slightly faster tempo that Bergman describes as "atos," all tempos at once)—they convey more of a sense that somebody might be out there listening. These ballads are almost tactile in their intimacy. Although their cragginess and tension save them from meditative prettiness, they'd sound like something on Windham Hill or ECM with just a little more reverb.

"God forbid I should be on ECM!" Bergman said in mock horror when I told him I was glad he wasn't. "I might sell more records and make some money. People have always asked me how I can play so fast, and now I can play slower than anyone else, too. I always wrote ballads, and maybe I would have played them if I had a little neo-bebop quartet. But playing them solo, they made me feel depressed, vulnerable. The problem was that I couldn't play them with as much drive as the faster pieces. Now I can, because I can now do something I couldn't do before, or at least not with ease, which is to play crossed hands continually, sometimes with the left hand in the treble and the right hand in the bass. Since the hands are mirror images, all the chords are inverted. Playing a ballad now, I feel exquisite, a little depressed, out of this world. When I play fast, my brain is working, along with my desire for identity or survival or whatever. When I play a ballad, I feel the sense of elegance you get when expressing a complete thought.

"So I have a serious dilemma, and I don't know what to do about it. My ballads I know I have an audience for. Not a mass audience, but a certain number of people who wouldn't like anything else I do. People *like* my ballads. But once you start seeking acceptance, you become less rambunctious. You take smaller steps."

A few days after I interviewed him, Bergman called me to ask me to recommend duet partners. This surprised me, because I think of him as a solo performer, hammering away at the keyboard and his thoughts ("making the silence answer," as he once put it). He also dropped me several notes, handwritten on unlined white paper, with parenthetical thoughts written vertically in the margins. These are examples of what he had to say:

"Francis, the hands are merely stooges now for my *ideas*." [underlined]

"My next step will be MEDIUM TEMPO and 'swinging' pulse—but now the left hand can go like a right—relaxed—THIS MAY enable me to BEAT THE ODDS— it's not Wynton Kelly, Red Garland, etc. but thinking of them playing 'Upside Down' _____ or Bill Evans, etc."

"I'll be able to play Albert Ayler with my left hand—that pathos."

"If someone asked me I'm playing BLACK and WHITE music: of course the piano has BLACK and WHITE keys—joke."

"The concept of dialogue will be based on the ability of each hand to play (alone) conceptions such as Circles, Whirl + Figure 8s, Swift River, Webs and Whirlpools, Uncharted Rivers, etc."

"I'm on to some kind of 'ecstacy'—it started with the voicings on 'Poignant Dream' (the improvisation) and 'Child Alone.'"

"I've always liked the Chicago players—there's a record on Columbia Bud Freeman, Condon, Kaminsky, Teagarden, Russell?, "After Awhile" "Muskrat Ramble" and others. The record you gave me [of Steve Lacy's early adventures in dixieland with the trumpeter Dick Sutton], fun, but diluted dixieland—fun, but not like the depth, poignancy, 'pain' + 'joy' of the earlier stuff."

[On the telephone] "I think those white Chicago musicians will eventually be considered significant for epitomizing a consciousness that existed at the time. it wasn't the robust, ebullient, wonderful thing that came from New Orleans. It was different: to be alive and drinking and playing jazz. I must start listening to the thirties again."

"I don't know what I'll do now—I'm at a crossroads—in two weeks I'll know— I'm getting out of teaching in June, if I keep the Steinway B—if I sell it I can get out now."

"Someday I'll tell you the REAL story of my life—I TRUST you'll be amazed."

(*October, 1987*)

OUTCHICKS

The soprano saxophonist Jane Ira Bloom holds a master's degree in saxophone, but her approach to music is intuitive rather than analytical, as witnessed by her choice of college. She could have attended the Berklee School of Music or the New England Conservatory, both located in Boston, close to Newton, the comfortable suburb where she was born. "But I went to Yale because I sensed that an isolated musical education wasn't the right choice for me. I felt that other interests were good, and that I would continue to make music wherever I was. I've been consumed by music since I was four years old: my mother had a bit of a record collection, and her Ella Fitzgerald albums put the sound in my ear."

The pianist Michele Rosewoman is Bloom's opposite in terms of formal education. Rosewoman says that she learned music "in the streets and clubs" of her native Oakland, California, where her parents ran a small record store specializing in jazz and Third World music. But Rosewoman, who talks black, looks polyethnic, and says only that "my background, going way back, is Spanish, but not Latin American," approached the task of self-education with a determination that would put most doctoral candidates to shame. "There was a four- or five-year period when I was involved with music twenty-four hours a day. I hardly slept, and when I did, I dreamed about music and rolled out of bed in the mornings still thinking about it. When I wasn't playing, I was listening."

Rosewoman enrolled as a music major at the University of California in Berkeley to please her mother, an alumna of that school, but dropped out midway through her freshman year when she realized that "there was nothing in the curriculum, which mostly dealt with fifteenth- to nineteenth-century harmony, that was going to be of much use to me. I wasn't even accepted as a music major at first, because I improvised at my audition, and they didn't like that. Meanwhile, I started hanging out at Laney Junior College, where Ed Kelly was teaching and leading a big band. He's a great pianist known mostly around the Bay Area. He's the only person I would point to as a direct influence, although I also listened carefully, in my formative years, to Coltrane, Monk, Lee Morgan, and Miles Davis's group with Wayne Shorter and Herbie Hancock. At the time I met Ed, I didn't know how to read a big band chart, but I auditioned for his band and made it just improvising on the blues. Back at UC, I began asking questions about the things I was learning playing with Ed's big band, and I was told, for example, that there was no such thing as a diminished scale. And when I asked about fourth chords and things of that nature, I became the brunt of all the jokes in the class. That's when I disappeared from there."

What Bloom and Rosewoman have in common, aside from both being in their early thirties, is the covert discrimination they've encountered as women instrumentalists (covert because, as Bloom points out, there's no way of counting how many calls for work you don't get). Male jazz musicians sometimes behave as they believe a woman's place is in the wings—a baffling attitude for a subculture that mocks most taboos. Given the music's black orientation, this raises the troubling question of whether black men in general are more chauvinistic. As Linda Dahl pointed out in *Stormy Weather: The Music and Lives of a Century of Jazz Women* (Pantheon, 1984), women have participated in jazz from the beginning, working alongside men on plantations, and presumably joining in field hollers. And it's impossible to overestimate the formative role played by the matriarchs of the black church. But by and large, women in jazz have been limited to exercising what Eldridge Cleaver called "pussy power": sleeping with righteous dudes and bearing their offspring. As jazz ceases to be a moneymaking proposition, wifely duties sometimes also include managing a husband's or lover's business affairs and supplementing his earnings by holding down a nine-to-five job. Though women have always been represented on the bandstand as singers, they've sometimes been merely tolerated in this capacity, as a necessary concession to mass taste. What would a

Freudian make of the fact that most of the handful of women instrumentalists who have achieved recognition from their male peers have played piano—an instrument you sit down to play?

Despite the belated elevation of Mary Lou Williams to pantheon status, the gradual acceptance of Carla Bley and Toshiko Akiyoshi as important composers, and the quick recognition given such relative newcomers as Bloom and Rosewoman—all within the last decade or so—women instrumentalists remain so much of a novelty that a recent ad for a Philadelphia appearance by the tenor saxophonist Joe Henderson emphasized that he would be fronting an "all-girl" rhythm section. Forget the indignity of "girl"—Can you imagine a club advertising an "all-black" rhythm section in support of a white leader? After arriving in New York in 1978, Rosewoman was told that as a result of this novelty factor, "I would get all the breaks, but that simply hasn't happened. I've had to prove myself beyond what a man would have to prove."

"I can pick up sexist overtones in the way music is described in print," says Bloom, who now lives in New York with her husband, Joe Grafasi, an actor (on screen in *Ironweed* and *Moonstruck*, and on stage in Tom Griffin's Off-Broadway hit *The Boys Next Door*). "If a woman is playing, music is described passionately as—oh, goodness, sweeping swerves of orgasmic sound." She responded with a hearty laugh when read a statement by a male saxophonist who explained his "love relationship" with his tenor by comparing the baritone to "a woman [who weighs] 300 pounds, [who's] demanding and overpowering," the alto to "a woman you can just overpower," and the soprano to "a woman who's totally insane."

If men see their horns as mates to be fondled into submission, what is a horn to a woman? "Whew! I can't say I have that kind of graphic sexual attachment to it. It's simply my voice. It's part of me. Playing it becomes as natural as breathing, and I almost forget it's there."

"Everybody has a bad night now and then," Rosewoman told me. "But if a young male saxophonist that everyone's read about is having a bad night, people just assume that's the case and say, 'I'll come back, hear him another time before making up my mind.' But with a woman, it just confirms what they expected of you going in. I've also had to resolve not to get involved in a man-woman thing with any of the musicians I've worked with—not to put myself in a situation where my femininity would be held against me. When I first began to play, I had to fight the assumption that, because I was a woman, I could only play delicately, not STRONG. That used to baffle me, because I was raised in a nonsexist environment, with no preconceived notions of what being a woman meant. I thought that was one part of what I was supposed to discover. I went through a period of looking at the world through the eyes of someone who had discovered what it meant to be oppressed as a woman in this society, before I did a turnaround and decided that it wasn't going to do me any good to see the world that way."

Rosewoman explains that her unusual surname "grew organically out of the name I had at birth, and I don't know that I want to talk about it. It's not that important. It was a nickname that stuck, and now it feels like my name. It's easy to remember. Musicians sometimes call me Rosewoman, instead of Michele." Rosewoman was briefly active in "woman's music" while still on the West Coast. (In fact, the all-woman jazz-rock band Alive! grew out of the informal workshops she gave in Haight Ashbury in the mid-seventies.) "But there wasn't much of an audience for instrumental music, especially jazz, in that scene," she says of a movement characterized by the lightheaded message-pop released on women-owned labels like Olivia and Redwood. "The women involved in that scene wanted to hear music with lyrics that reflected social or political views. Plus, it's a very undiscriminating audience. Whatever a woman gets up on stage and does, that audience loves her to death, and that's not much of a challenge for a musician. There's a great singer named Linda Tillery who's been hurt, number one, because she's been exposed only to the women's music community, and, number two, because she hasn't been challenged."

On *Modern Drama* (Columbia FC-40755)—Bloom's fifth album, but her first for a major U.S. label—she unveils an electronic device she calls "the gizmo," for lack of a better name, designed for her by Kent McLagin, a bassist and electrical engineer. "I sometimes whip my horn around in a 180-degree arc as I play, and for years I've been interested in how movement affects sound. So Kent and I came up with a small device that attaches to the bell of the horn, measures the velocity, and translates it into an actual change in timbre. The end result, you could say, is a silvery, shimmery kind of sound. It's just a prototype. I'm not traveling with it, because it's not roadworthy yet, and it's cumbersome because it's attached by a small wire to a couple of boxes that look like something out of a 1950s science-fiction movie."

She puts the gizmo to best use over the pianist Fred Hersch's doo-wop-ballad triplets on "Rapture of the Flat" and on "The Race," a breakneck track dedicated to race-car driver Shirley Muldowney. "I'm open to a lot of nonmusical influences, and I've found some of my role models among women like Shirley who have excelled against the odds. The combination of speed and grace fascinates me. I've been invited to join what's officially called the NASA Art Team. In the past, only visual artists have been invited to the Cape for the three or four days prior to a launch. But I've been asked to write a piece of music based on my impressions of a launch. I'm thrilled, because I've mused a lot about the space program, or more specifically, about the future of music in space. Did you know that Ron McNair, one of the astronauts who died in the Challenger, played soprano? I suppose he was just a hobbyist, because I don't know how anyone who was an astronaut would have much time for anything else. But he took his horn on the Challenger with him."

Though Bloom's approach to electronics is uncommonly "saxophonistic—musicians who work in the most sophisticated level of electronics would consider what I'm doing pretty primitive," the slightly worried tone she coaxes out of the soprano is so buoyant and personal that you wonder why she wants to distort it. *Modern Drama*'s biggest pluses are the gizmo-free ballads "Strangely and Completely" and "More Than Sinatra" (she wouldn't elaborate on that title). "Because like most musicians, I'm forever looking for new ways of making sound. I want to achieve a blend of what I've already put into my instrument and the new sounds that electronics makes possible. Soprano has never been a doubling instrument for me, even though I used to also play alto and studied tenor. Starting out, I wasn't in touch with the soprano saxophone lineage: Sidney Bechet, Steve Lacy, John Coltrane, Wayne Shorter. As a high school student, I studied with Joseph Viola, a master saxophone teacher from the New England Conservatory, who has a special affinity for the soprano which he passed on to me. I paid more attention to trumpeters and singers. The soprano is very direct-sounding, like a voice, and you have to finesse it, just like you do a trumpet, in order to get it to sing. I like Booker Little for his use of very large, very unusual intervals, and his strong sense of melody, the way he could communicate all sorts of things with the sound of one note; and Miles Davis for the range of his work, the way he epitomizes the constantly changing artist, the way he can make you cry with one note. Abbey Lincoln's influence is there, too—the subtlety of her phrasing, the depth and poetry and grit and something a little bit dark in her voice.

"My father is a camp director, and my mother takes care of the books for him. It must have made them nervous when I moved to New York to pursue a career in music, but they were very supportive. In New York, I had a few lessons with [tenor saxophonist] George Coleman. I sought him out because I wanted to know more about the sort of fast harmonic motion he's a master of—because his knowledge of harmony is so complete; he picks notes that are extraordinary; and he does it at tempos that would frighten most saxophonists. George was concerned for me. He worried that the insensitivity of club owners and record producers might be something I wasn't ready for, and he would sometimes go on jobs with me to make sure that I got paid what was coming to me. I've done very little work as a side person, and it's difficult to say whether that's been by choice. When I came to the city, I was open to anything that was offered to me, but not a whole lot of opportunities presented themselves—whether because I was a woman, I honestly don't know. But it was good, in one way, because it gave me the motivation to write music and form my own bands."

In addition to the light-textured front line blend of Greg Osby's soprano and Steve Coleman's alto, Rosewoman's *Quintessence* (Enja 5039) is notable for a rhythmic vitality stemming, Rosewoman says, from her research into Cuban folkloric music, to which she was first exposed as a teenager. "The approach to rhythm isn't a jazz approach. In 'For Now and Forever,' for example, the band, including [drummer] Terri Lyne Carrington, is playing on the upbeat. Only the solos introduce a downbeat feeling. 'Vamp for Ochun,' the funky track named for Ochun, a female deity in the Nigerian pantheon—people assume it's in an odd time signature, but it's in $\frac{4}{4}$. It grew out of my appreciation of what Afro-Cuban *bata* drummers can do to rhythm with their complex syncopations."

In terms of Rosewoman's own playing, the most revealing track in the boppish "Lyons." An homage to the late alto saxophonist Jimmy Lyons, the piece illustrates how much Rosewoman learned from Lyons' longtime compatriate, Cecil Taylor. Hearing Taylor's group with Lyons in San Francisco in 1972 was "the turning point for me," Rosewoman says. "Philip Elwood, a Bay Area jazz critic, arranged for me to get in free several nights to hear Cecil at the Keystone Korner. This was 1971 or 1972, and I wasn't familiar with his music. Before hearing Cecil, I thought of myself as a young player learning the tradition. I wasn't interested in the so-called avant-garde. At the time, I was in the process of moving, and didn't have access to a piano. When I did again, I found that my playing had changed, and I was able to trace that change to Cecil. Years later, I asked Philip Elwood why he had gotten me in to hear Cecil, and he told me that he heard something in my playing that indicated that I was heading in his direction. I went through an emotional and physical journey in those three or four nights. I was already playing percussively, but Cecil opened my head up to the concept of music as sound. I watched him to see where his resting spots were, because it wasn't immediately apparent. I wanted to understand the impulse that brought the sound out—how he heard what he heard. I remember being struck by Jimmy Lyons' poise, the peacefulness he projected, the way he rode the wave of Cecil's music, in contrast to [the trumpeter] Raphe Malik, who the music went right through, the way it went right through me."

When Rosewoman first arrived in New York, she earned her living as a piano tuner. She still gets calls in that capacity, and combined with her current job teaching music to school-age children for the New York City Parks Department, this affords her a certain independence. "I have health coverage and vacation pay, things that are luxuries for a musician. Plus, I like teaching. I turn down some jobs because I'm too busy." She hopes that her next album will mark the belated record debut of New Yuruba, the fourteen-piece band with three *bata* drummers ("traditionally, there are always three") which with she's performed in New York for the last several years. "My introduction to Cuban music was through the religious music, rather than the secular, and that folkloric drumming really tapped something in me." She describes New Yuruba as reconciling two influences on her music by "bringing two musical worlds together, contemporary jazz and Afro-Cuban religious music, an experimental approach on one side and a folkloric tradition on the other. It's like having everybody in my extended family together in one place."

Bloom and Rosewoman move in different musical circles, but their paths crossed briefly when they assembled, along with Carrington, the French horn player and pianist Sharon Freeman, and the pianists Geri Allen and Amina Claudine Myers, for a group photo for an article on women in jazz in *New York Woman.* "It was so nice to be in the same room with all those talented ladies—such a rare experience," Bloom says.

Did they talk about the problems they face as women in a male-dominated art form? "Nah, are you kidding?" Bloom laughs. "But we did talk about what we were up to professionally, and I was heartened by the fact that Geri and Terri Lyne are going out on the road with Wayne Shorter. That's a sign that things are changing. I'm on a very different path, leading my own group, but it's nice to know that there are women who are being offered the situations their abilities warrant."

(February, 1988)

eight singers
and a comic

THE MAN WHO DANCED
WITH BILLIE HOLIDAY

Since her death in 1959, Billie Holiday has achieved a sainted second existence in our national imagination. In death as in life, she is hailed as the greatest woman jazz singer of them all, but her legend transcends music, just as it transcended mortal bounds. Holiday is even more a pop icon today than during her lifetime—regrettably, less as a result of the reissues that have proliferated in recent years than as a side effect of *Lady Sings the Blues*, the 1972 Diana Ross star vehicle ostensibly based on the itself-none-too-reliable 1956 autobiography Holiday co-authored with William Dufty (Doubleday). In addition to her own colorful account, there have been at least two other Holiday biographies— Alexis DeVeaux's *Don't Explain* (Harper and Row, 1980), a shrill book-length prose poem that mother-milks the injustices Holiday suffered as a black woman for cheap pathos, and John Chilton's *Billie's Blues* (Stein and Day, 1975), a workmanlike sorting out of hard facts which limits itself to Holiday's twenty-five-year recording career.

But the Billie Holiday bibliography isn't limited to the books that inquire into her personal life or those that analyze her contributions to jazz. She is a recurring apparition in novels, poems, plays, and literary memoirs, including *The Autobiography of Malcolm X* (Grove Press, 1965) and Frank O'Hara's "The Day Lady Died" (*Lunch Poems*, City Lights Books, 1964). Her independence and the devastating costs, both real and psychological, that she paid for it make her a creature of premonitory fascination for women writers in particular.

> Billie is late—of course, she is always late—but the crowd in this packed room is not resigned: people are restless, and tense. There is a lot of lighting of

cigarets, looking at watches, loud orders for more drinks. And there are scattered rumors: she's sick, she's not coming, been in a wreck—she's just phoned to say she'll be there in ten minutes. And eerily, throughout all that waiting for Billie, on the huge and garish jukebox one of her records is playing; from out of all that poison-colored neon tubing Billie's beautiful, rich and lonely voice is singing, "I cover the waterfront, I'm watching the sea—"

And then suddenly she is there, and everybody knows, and they crane their heads backward to see her, since she has come in by the street entrance like anyone else. Or, not like anyone else at all: she is more beautiful, more shining, holding her face forward like a flower, bright-eyed and smiling, high cheekbones, white teeth and cream-white gardenia at her ear. . . .

With a wonderful gesture Billie throws her coat down on the stage, and for a moment she stands there in the spotlight, mouthing the words coming from the jukebox—"Will the one I love, be coming back to me?"—as everyone laughs and screams and applauds.

Somewhere in that audience, probably up near the front, is a very young and pretty small girl, who is not paying much attention to Billie. Eliza Hamilton, with long smooth blond hair that curls suddenly at the ends, and dark blue eyes. She has serious and obsessive problems of her own: is she pregnant? Her heavy breasts are heavier, and sore. And if she is pregnant, what should she do? Should she marry Evan Quarles, the paler blond and sad, Deep Southern young man at her right? He would like them to marry, and that is strange: Eliza knows that she is more in love than he is, but it is he who urges marriage. He is deeply disturbing, mysterious to her; she is both excited and obscurely alarmed by Evan—is that "in love"? . . . Now she looks at Evan with a mixture of enmity and curiosity: who *is* he?

Eliza is barely listening to Billie, who now, with her small combo in the background, is singing, "Once they called it jazztime, to a buck and wing—"

Singing, swinging it out.

But Eliza retained that scene of Billie's entrance, and Billie singing. (Singing what? What was she wearing?) She kept it somewhere in her mind; she brought it out and stared at it as she might a stone, something opalescent. At times she wondered how much of it she had imagined. . . .

Should she have an abortion? Who would know a doctor who would do it? Who would pay for it? . . .

But even in the midst of such frenzied speculation, Eliza is aware that this evening—these hours—are important; she *knows* she will remember. And she thinks of the following Monday, when she will be back in Connecticut, in school, and she will tell her friends about seeing Billie—how beautiful she was, her voice. How Evan Quarles, the interesting older man, took her to Fifty-second Street to hear Billie Holiday. She will not tell any friend that she might be pregnant.

Billie has stopped singing, left the stage, and Evan says, "I'd buy you a gardenia if I weren't allergic to them." . . .

Will he want to make love later? Will he take her back to his place, on Horatio Street, in the Village? Eliza can't be sure of anything with Evan. . . .

Eliza looks around at the other talking people, and she suddenly perceives, feels, that there is an extraordinary number of handsome young men, all strangers, all unexplored and possible. She looks at them intently . . . so *atractive* all of them. Aware of her own look, its intentness, she wonders what message she is delivering: is she somehow inviting them, or saying goodbye, as she would to other men if she should marry? And if she should not be pregnant, will she meet one of these new young men months later, and together will they remember hearing, seeing Billie? Will that happen?

Then Eliza notices that the young woman at the next table is heavily pregnant, so huge she must sit back in her chair. Eliza's spirits sink, her fantasy vanishes. She recognizes that young woman as an omen, a terrible sign: she, Eliza, *is* pregnant. . . .

Now Billie is walking back onto the stage; amid thundering applause, shouts, and whistles, she saunters into the smoke-beamed center of light; she stands there, one hip thrust forward. She scans the crowd as though she could see everyone there. Is she possibly seeing the men and feeling the urgent attraction that Eliza felt a few minutes before? Her eyes are blank, and her smile says nothing.

"She looks bad," Evans whispers—too loudly, Eliza feels, even in this noisy room. "Drugs—she can't last long."

"Georgia, Georgia, no peace I find . . ." sings Billie, whose beautiful face has come alive, whose eyes say everything.

Alice Adams, *Listening to Billie* (Alfred Knopf, 1977)

I had heard stories of Billie being beaten by men, cheated by drug pushers and hounded by narcotics agents, still I thought she was the most paranoid person I had ever met.

"Don't you have any friends? People you can trust?"

She jerked her body toward me. "Of course I have friends. Good friends. A person who don't have friends might as well be dead." She had relaxed, but my question put her abruptly on the defense again. I was wondering how to put her at ease. I heard Guy's footsteps on the stairs.

"My son is coming home."

"Oh. Shit. How old you say he is?"

"He's twelve and a very nice person" . . .

"Billie Holiday? Oh. Yes. I know about you. Good afternoon, Miss Holiday." He walked over and stuck out his hand. "I'm happy to know you. I read about you in a magazine. They said the police had been giving you a hard time. And that you've had a very hard life. Is that true? What did they do to you? Is there anything you can do back? I mean, sue them or anything?" . . .

Billie's face was a map of astonishment. After a moment, she looked at me. "Damn. He's something, ain't he? Smart. What's he want to be?"

"Sometimes a doctor, and sometimes a fireman. It depends on the day you ask him."

"Good. Don't let him go into show business. Black men in show business is bad news. When they can't get as far as they deserve, they start taking it out on their women. . . ."

. . . She stayed for dinner, saying that I could drop her off on my way to work. She talked to Guy while I cooked. Surprisingly, he sat quiet, listening as she spoke of Southern towns, police, agents, good musicians, and mean men she had known. She carefully avoided profanity and each time she slipped, she'd excuse herself to Guy, saying, "It's just another bad habit I got." After dinner, when the baby sitter arrived, Billie told Guy that she was going to sing him a good-night song.

They went into his room, and I followed. Guy sat on the side of his bed and Billy began, a cappella, "You're My Thrill," an old song heavy with sensuous meaning. She sang as if she was starved for sex and only the boy, looking at her out of bored young eyes, could give her satisfaction. . . .

For the next four days, Billie came to my house in the early mornings, talked all day long and sang a bedtime song to Guy, and stayed until I went to work. She said I was restful to be around because I was so goddam square. Although she continued to curse in Guy's absence, when he walked into the house her language not only changed, she made considerable effort to form her words with distinction.

On the night before she was leaving for New York, she told Guy she was going to sing "Strange Fruit" as her last song. We sat at the dining room table while Guy stood in the doorway.

Billie talked and sang in a hoarse, dry tone the well-known protest song. Her rasping voice and phrasing literally enchanted me. I saw the black bodies hanging from Southern trees. I saw the lynch victims' blood glide from the leaves down the trunks and onto the roots.

Guy interrupted, "How can there be blood at the root?" I made a hard face and warned him, "Shut up, Guy, just listen." Billie had continued under the interruption, her voice vibrating over harsh edges.

She painted a picture of a lovely land, pastoral and bucolic, then added eyes bulged and mouths twisted, onto the Southern landscape.

Guy broke into her song. "What's a pastoral scene, Miss Holiday?" Billie looked up suddenly and studied Guy for a second. Her face became cruel, and when she spoke her voice was scornful. "It means when the crackers are killing the niggers. It means when they take a little nigger like you and snatch off his nuts and shove them down his goddam throat. That's what it means."

The thrust of rage repelled Guy and stunned me.

Billie continued, "That's what they do. That's a goddam pastoral scene."

Guy gave us both a frozen look and said, "Excuse me, I'm going to bed." He turned and walked away.

I lied and said it was time for me to go to work. Billie didn't hear either statement.

I went to Guy's room and apologized to him for Billie's behavior. He smiled sarcastically as if I had been the one who had shouted at him, and he offered a cool cheek for my good night kiss.

Maya Angelou, *The Heart of a Woman* (Random House, 1981)

She was fat the first time we saw her, large, brilliantly beautiful, fat. She seemed for this moment that never again returned to be almost a matron, someone real and sensible who carried money to the bank, signed papers, had curtains made to match, dresses hung and shoes in pairs, gold and silver, black and white, ready. What a strange, betraying apparition that was, madness, because never was any woman less a wife or mother, less attached; not even a daughter could she easily appear to be. Little called to mind the pitiful sweetness of a young girl. No, she was glittering, somber and solitary, although of course never alone, never. Stately, sinister, and determined.

The creamy lips, the oily eyelids, the violent perfume—and in her voice the tropical *l*s and *r*s. Her presence, her singing created a large, swelling anxiety. Long red fingernails and the sound of electrified guitars. Here was a woman who had never been a Christian.

To speak as part of the white audience of "knowing" this baroque and puzzling phantom is an immoderation and yet there are many persons who have little splinters of memory that seem to have been *personal*. At times they have remembered an exchange of some sort. And of course the lascivious gardenias, worn like a large, white, beautiful ear, the heavy laugh, marvelous teeth, and the splendid head, archaic, as if washed up from the Aegean. Sometimes she dyed her hair red and the curls lay flat against her skull, like dried blood. . . .

In her presence on these bedraggled nights, nights when performers all over the world were smiling, dancing, or pretending to be a prince of antiquity, offering their acts to dead rooms, then it was impossible to escape the depths of her disbelief, to refuse the mean, horrible freedom of a savage speculation of destiny. And yet the heart always drew back from the power of her will and his engagement with disaster. An inclination bred from punishing experiences compelled her to live gregariously and without affections. . . .

A genuine nihilism; genuine, look twice. Infatuated glances saying, Beautiful black star, can you love me? The answer: No.

Somehow she had retrieved from darkness the miracle of pure style. That was it. Only a fool imagined that it was necessary to love a man, love anyone, love life. Her own people, those around her, feared her. And perhaps even she was often ashamed of the heavy weight of her own spirit, one never tempted to the relief of sentimentality. . . .

The sheer enormity of her vices. The outrageousness of them. For the grand destruction one must be worthy. Her ruthless talent and the opulent devastation. Onto the heaviest addiction to heroin, she piled up the rocks of her tomb with a prodigiousness of Scotch and brandy. She was never at any hour of the day or night free of these consumptions, never except when she was asleep. And there did not seem to be any pleading need to quit, to modify. . . .

She lived to be forty-four; or should it better be said that she died at forty-four. Of "enormous complications." Was it a long or a short life? The "highs" she sought with such concentration of course remained a mystery. I fault Jimmy for all that, someone said once in a taxi, naming her first husband, a fabulous Harlem club owner when she was young. . . .

Her whole life had taken place in the dark. The spotlight shone down on the black, hushed circle in a cafe; the moon slowly slid through the clouds. Night—working, smiling, in makeup, in long, slinky dresses, singing over and over, again and again. . . .

Elizabeth Hardwick, *Sleepless Nights* (Random House, 1979)

It isn't just in books that one encounters memories of and fantasies about Billie Holiday. Early one New Year's Eve, years ago, a tall, trim white man in his early sixties, who had visibly had many more than one too many, stumbled into the record store I was managing. A Billie Holiday record was playing—I don't remember which. He told me of another New Year's Eve, decades earlier, when as a sailor at liberty in New York, he had danced with Billie Holiday.

He was young then, he told me, and so was she. This supposedly was before she had recorded, before anyone knew who she was. She was performing unbilled in a Harlem nightclub where she and the other female performers were required to mingle with the male patrons and hustle drinks between sets. After hearing her sing, he decided he wanted to dance with her, even though he'd had so much to drink that it was an effort to stand—and even though he had never so much as spoken with a black woman before. At the urging of his shipmates, he found her alone at the bar and asked her to dance—an invitation she accepted silently, with no discernible enthusiasm, he thought.

As he struggled to keep his balance and she struggled to hold him up, he told her that she was the most beautiful woman he'd ever seen. She told him he was drunk. As they danced, he pressed a handful of bills into her palm; he made a point of shoving another handful down her cleavage. She said nothing, but once or twice he fancied that he felt her fingers trying to unlock his to engage them in a caress; once he thought he felt her hand brush against his groin. But he was so numb from the liquor that each time intimate contact was made—if it had been made—it was over before he could respond.

When the dance was over, his buddies cheered him loudly and helped him to his chair. He hollered for another round of drinks, on him, and, digging into his pocket to pay for them, discovered the six crumpled twenties that she had returned there. One hundred and twenty dollars, a good piece of change for the time.

He told me that he considered Billie Holiday's action (let's pretend for a moment that it was Billie Holiday) of rejecting both his money and his drunken advance an unquestionable and unsolicited kindness.

To say that Billie Holiday was a great jazz singer hardly says enough, for she defined the very role of the modern jazz vocalist much the way her early model Louis Armstrong had defined the role of the jazz soloist. Inspired by work songs and field hollers, spirituals and the blues, as well as by the shining brass of military bands, the seminal jazz instrumentalists sought to replicate the cry of the human voice. And the earliest jazz singers reversed the strategy, the hornlike sweep of their pendulous voices taking precedence over the meaning of the words they sang. As heiress to that tradition, Holiday was often applauded (and justly so) for her ability to phrase like a horn. Jimmy Rowles—one of her favorite and most frequent accompanists in the last decade of her life—says of her: "She had a great sense of humor, and you never had to mind your manners or your mouth when you were with her, because she could lay anybody out with *her* language when she wanted to. You could say anything you pleased around her—none of that ladylike crap. She was a regular guy, you might say." It's significant that in the next breath Rowles adds, "She never had any problem with chord changes, either." Even more than her readiness to cuss with the best of them, Holiday's unerring musicianship made her one of very few woman singers whom male instrumentalists considered enough their equal to drop their guard around.

On her earliest recordings with Teddy Wilson, we hear Billie Holiday obeying the custom of the day, lining up with the instrumental soloists, usually singing no more than one chorus in what were essentially instrumental performances. But with Holiday (and Mildred Bailey and Lee Wiley), jazz made the happy discovery that words could be of paramount importance. Such were Holiday's powers of evocation, her capacity to illuminate conflicting emotions with the slightest turn of phrase, that in her wake we expect the finest women jazz singers to be great dramatic actresses as well, and judge them harshly when they come up short. Just as Armstrong's emergence from the New Orleans ensemble signaled the rise of the improvising soloist, so Billie Holiday skyrocketed the jazz vocalist to spotlight prominence.

But those of us who pledge our allegiance to jazz have seldom been granted the luxury of keeping a great singer all to ourselves. Singers succumb to blandishments rarely offered their instrumental peers, why is why we've had to share so many of our favorites with lovers of soft lights and sweet music, and why it was probably inevitable that as sublime an interpreter of popular songs as Holiday would eventually forsake interactive small jazz groups for the tatty, clinquant luster of large, impersonal studio orchestras, as she did for a time (on records, at least) in the forties. But if Holiday was the most exquisite of singers, she was also the least pretentious. She was in her element when she could mix it up with a couple of horns, and it's to producer Norman Granz's everlasting credit that she again got the chance.

I doubt that it's still necessary for those of us who admire the records that Holiday recorded for Granz's Clef and Verve labels in the fifties to rally to their defense. Generally regarded as painful evidence of Holiday's decline when initially released, these albums are now cherished by some as Holiday's most triumphant, and you don't have to agree with that revisionist wisdom to concede its logic. By the time Holiday signed with Granz in 1952, the tomboy ebuillience that first endeared her to listeners had vanished, along with most of her timbral elasticity. But in compensation for a voice that had darkened and stiffened and sunk into the hollows, she could claim a ripened sensitivity to nuance that, especially on ballads, made her singing even more intimate, powerful, and affecting. Despite ever-constricting artistic and personal horizons, she stood her ground and sang without apology, which is why her records from the fifties hardly need apology three decades later.

A ravaged but still unvanquished Billie Holiday is reason enough to treasure her Verves, but there are a number of other factors in their favor. The selection of songs is worthy of a great singer, with a heavy emphasis on George and Ira Gershwin, Irving Berlin, Cole Porter, Rodgers and Hart, Mercer and Arlen, and the other venerables of American popular song—no novelties of the sort she had to contend with in the thirties (granted, she performed miracles on some of them), and no broad torch songs of the kind for which she had surprisingly little affinity. Another key to the success of the series was the sympathetic accompaniment Granz enlisted for her. "She didn't want to hear any banging around going on behind her," says Rowles, the pianist on many of her dates for Granz, including the five from January, 1957 included on *Embraceable You* (Verve 817-359-1), the fourth and final double volume of Holiday's studio recordings for Granz. "She was a lot like Lester Young in that respect. In fact, she measured everybody by the way Lester had played behind her years before."

Although Ben Webster's chesty appoggiatura put him at some remove from Young stylistically, he proved to be just as compatible a leading man for Holiday. According to Chilton's *Billie's Blues*, Webster and Holiday had been lovers back in the thirties, a brief but stormy affair that ended with Holiday nursing a black eye and her mother retaliating against Webster with an umbrella. By 1957, however, Holiday and Webster had put all such passion and acrimony behind them. They were old friends, no less and no more, but their musical partnership was one of those marriages made in heaven: the most musical of singers and the most lyrical, most vocal of tenor saxophone balladeers. If I prefer the Verve "Embraceable You" to the version that Holiday made for Commodore in 1944, and the Verve "Body and Soul" (also included here) to the more celebrated 1940 Okeh—and sometimes I think I do—it's not so much Holiday's greater maturity that tips the scale as Webster's smitten obbligatos and effulgent solos in support of her. There's something rakish and heroic about the way the Clark Gable of the tenor saxophone sweeps a pretty melody up in his embrace and sweet-talks it out to the heart of a crowded dance floor, something magnificent about his pleading ardor and persuasive finesse.

Holiday also went back a long way with the trumpeter Harry "Sweets" Edison, her teammate in the 1937 Count Basie big band, whose staggered attack and muted drolleries make him as effective a foil to Webster as to Holiday. Edison's sharp wit is displayed to good advantage in his playful exchanges with Holiday on "Just One of Those Things." The guitarist Barney Kessel (probably the only musician who can boast of being on records by Holiday, Fred Astaire, Charlie Parker, Sonny Rollins, and the Ronettes) provides shimmering pickups, intros, and solos that disguise their harmonic urbanity in engaging country twang. The bass chores are split between Red Mitchell and Joe Mondragon, with Alvin Stoller and Larry Bunker alternating on drums: firm and unobtrusive rhythm players, all of them. And finally, there is Jimmy Rowles, who demonstrates beautiful empathy with Holiday on their fragrant duet readings of the verses to "A Foggy Day" and "I Didn't Know What Time It Was," and their confident delineation of "Say It Isn't So."

"I met her for the first time in California in 1942," Rowles remembers. "I loved her immediately, and after she looked *me* over once or twice, we got along fine. She was old friends with you immediately if she liked you. If she couldn't stand you or thought you were a phony, she'd let you know that right away, too. I was never really a member of her band. I never went on the road with her, but she'd ask for me whenever she came out to work in Los Angeles.

"The dates we did for Verve always started around two-thirty in the afternoon, and the atmosphere would be totally relaxed. There was always plenty to drink, and if anybody wanted sandwiches, Norman would send out for some. Norman would be in the control room with a long list of tunes, and Billie and I would get together to settle on keys. What key she'd sing a number in would depend on what she'd been doing the night before. If she'd been out celebrating all night, her voice would be even gravellier than it was to begin with, so we'd take it in whatever key she could make. We got into some mighty weird keys that way, because of the ups and downs of her voice. There would never be many takes. She wouldn't stand for more than two or three. If something went wrong, we'd stop and figure it out, say 'I'll do this while you do that,' and away we'd go. It was like a jam session. Norman's Granz's head is one big jam session anyway, bless his heart."

The inclusion of the previously unreleased "Just Friends"—a Holiday-less jam with relaxed solos by Rowles, Kessel, Webster, and Edison—comes as a complete surprise to the pianist, who was working around Los Angeles with Webster and Edison at the time of the sessions. "Well, I'll be damned. I don't even remember us doing it, and I don't know how it could have come about. I wonder where Billie was? I don't remember Billie ever being late or taking any longer breaks."

According to Rowles, Holiday was in good spirits at these sessions; an impression confirmed by the optimistic tilt of her phrasing, even on the ballads. Holiday's star again seemed to be on the rise in the early months of 1957. The fanfare that greeted the publication of Lady Sings the Blues the previous summer climaxed with an emotional career retrospective at Carnegie Hall in November, and rumor circulated that Hollywood was interested in purchasing the rights to Holiday's story as a vehicle for Dorothy Dandridge. In the sanguine flush of all this publicity, Holiday announced plans to sue the New York City Police Department for the cabaret card so long denied her. For whatever combination of reasons, however, these sessions with Webster and Rowles turned out to be her last recorded hurrah. She made just one more record for Granz, a spiritless live set from the 1957 Newport Jazz Festival. After taping "Fine and Mellow" for the CBS-TV special The Sound of Jazz, with an entourage of great saxophonists (Webster, Young, Coleman Hawkins, and Gerry Mulligan) in the spring of 1957, she entered the recording studio only twice more, both times shrouded in sepulchral string arrangements. And less than three years later, the brazen life she led finally claimed her martyr.

A quarter century later, the memory of Billie Holiday—the burning need to remember her—still haunts us. Whether we like admitting it or not, Holiday's metaphorical afterlife owes something to the American appetite for the sentimental and the sensational, the need we have as a people for constant reassurance that the meek shall inherit the earth and the wages of sin is death. But the grim statistics of one woman's troubled existence, the masochistic squalor of her demise, would have drifted away into the ether by now if not for the ballast of her universal, intimate, and singular art. What we remember when we think we remember seeing Billie Holiday, talking with Billie Holiday, even touching Billie Holiday, is listening to her. And as long as we have access to her on records as glorious as *Embraceable You*, she will continue to be the most persistent of memories, and much more. Much, much more.

(1983–1985)

ELLA

The just re-released version of George Cukor's backscreen musical *A Star Is Born* begins with previously excised footage of Judy Garland's Vickie Lester torching Harold Arlen and Ira Gershwin's "The Man That Got Away" at a jam session, as James Mason's Norman Maine vamps her from the shadows. Reluctant to perform at first, Garland stalks the screen and sings as if possessed, unaware of Mason's predatory eye, seemingly unaware even of her accompanists, whose horns point heavenward in commiseration with her lament. The entire sequence is electrifying, with slow, almost hallucinatory camera movements. But it had to be reshot, presumably because Garland looked dowdy and truculent in the rushes; her weight fluctuates throughout the long film, and so does her mood—which is ironic, because Mason is the one who's supposed to be cracking up. In the take moviegoers saw in 1954 (and we see it, too, later, in proper chronological sequence), Garland's singing seems less convincing because Cukor's visuals aren't as deliriously stylized. Still, when Mason tells Garland that he wants to groom her for stardom—describing his reaction to her performance in terms of "little jabs of pleasure" and "jolts of electricity" and comparing her to a prizefighter "sniffing blood and moving in for the kill"—we know what he's talking about, not having quite recovered from the opening insert ourselves.

I doubt anyone would use such carnal imagery to describe Ella Fitzgerald's girlish, unshadowed reading of "The Man That Got Away," which she recorded in 1961 for her just-reissued two-record *Harold Arlen Songbook* (Verve 817-526-1). Unlike Garland or Billie Holiday, Fitzgerald is a presentational singer who imposes no subjective weight on her material and unlocks no secret, ambivalent yearnings in herself or her audiences. She is no wild Duse, and no diva, either. Songs rarely become mere props for her virtuosity, as they often do for Sarah Vaughan's. Even at her best, Fitzgerald offers no pleasures deeper than those of hearing a good song superbly sung. But her genius is that she usually manages to make this seem like enough.

Even more than her free-flying scat, what we marvel at in Fitzgerald is her ability to disappear in words and melody without leaving a trace. The most beguiling demonstration of this remains the durable *Songbook* series she recorded for Verve beginning in the late fifties, and the Arlen set is the pick of the bunch, possibly because Arlen was a fine jazz singer himself early in his career and had the good fortune to collaborate with lyricists (like Ted Koehler, Yip Harburg, Johnny Mercer, and Ira Gershwin) who could hook his melodies to everyday figures of speech that jibe wonderfully with Fitzgerald's matter-of-fact delivery. There's a delightfully vulgar bounce to Arlen's songs—an urban snarl beneath their veneer of wit and sophistication—that helps them resist the sacred-text treatment favored by American Popular Song epigones (the sort of people who insist on the upper case). Arlen's melodies are ideal for Fitzgerald, who exploits their full improvisatory potential merely by holding fast on their hairpin turns. The brassiness of Billy May's horn charts underlines both Arlen's enduring vitality and Fitzgerald's innate swing; even his string arrangements have a sparkle that suits the material. May and Fitzgerald excel at staging fullblown Arlen production numbers, like "The Man That Got Away," "Stormy Weather," "Over the Rainbow," and "Blues in the Night." But the most engaging Arlen songs (and the most fruitful Fitzgerald and May interpretations) are those that sound tossed-off: "Let's Fall in Love," "As Long as I Live," or, best of all, the previously unissued "Sing, My Heart," which finds Fitzgerald riffing on vowels like a saxophonist executing a precipitous scaler run. The solos of Ben Webster and Stuff Smith make *The Duke Ellington Songbook, Volume Two* (Verve VE2-2540) a superior jazz showcase, but the songs are better here (though he was a great composer, Ellington was no match for Arlen as a songwriter—nobody was). Moreover, on the *Arlen Songbook*, the brief interludes by the alto saxophonist Benny Carter, the tenor saxophonist Plas Johnson, and the trumpeter Don Faggerquist, though infrequent, are worth waiting for. And though Fitzgerald's five-record Gershwin box (Verve 2615-063) may be the more definitive composer's retrospective, the overripe colors of Nelson Riddle's arrangements sound faded next to May's deft, workmanlike sketches for the Arlen tunes.

Fitzgerald's Songbooks were such masterpieces of their kind that encores were inevitable. I wish I could hail *Nice Work If You Can Get It* (Pablo D2312-140), her newest collection of Gershwin, as a latter-day triumph. Failing that, I wish I could pin all its shortcomings on the slumming lapses of her accomplice, the pianist André Previn (was doing "Let's Call the Whole Thing Off" in three/four time his bright idea?) But Fitzgerald is the real culprit. Not only is her voice not what it used to be (that much is forgivable), her adherence to melody isn't what *it* used to be, either; and her uvular gymnastics risk self-parody (on "A Fog-gog-gog-ey Day," in particular). Previn's rococo touches hardly encourage the singer to toe the line. Still, he and the world-class bassist Niels-Henning Orsted Pederson are frequently right on the money. *Nice Work If You Can Get It* brims with felicities, including a brisk, efficient "Who Cares?" and an understated reading of "How Long Has This Been Going On?" on which Fitzgerald expertly registers a playgirl's awed first brush with true love as Previn splashes ambiguous dissonances behind her. But even on these tracks, the overfamiliar material works against the singer, who has, after all, recorded impeccable versions of these songs in the past.

At her best, Fitzgerald inhabits a song with her total being, yet leaves it much as she found it—a feat more difficult and gratifying than it sounds, like a novelist holding the proverbial mirror up to life without leaving his messy fingerprints all over the glass. In her selfless interpretations, the works of the great American songwriters become a colloquial poetry that speaks both to us and for us. Which is not to say that standards are superior to rock tunes, only that like Garland, Holiday, or Sinatra, rock singers and songwriters have been trained to go for the kill. When they succeed, it can trigger revelation, but when they fail, you want to echo Gatsby's dismissal of Daisy's feelings for Tom: it was only personal.

(June, 1984)

A MAN AND
HIS MISHEGOSS

Not so much a boxed set as six different Frank Sinatra retrospectives in one slipcase, *The Voice: The Columbia Years, 1943–1952* (Columbia C6X 40343) attempts a rehabilitation project as difficult as the one that faced Sinatra himself immediately after the termination of his contract with the label in question. Unlike his early RCAs with Tommy Dorsey or his mature Capitol albums of the mid-fifties, Sinatra's Columbia sides have never been prized—though the best of them, all present and accounted for on *The Voice*, deserved to be. The Columbias have an aura of failure antithetical to the Sinatra mystique; his star, still ascendant when he signed with Columbia as a solo performer in 1943, had plummeted a decade later. *The Voice* makes it obvious that this wasn't a result of artistic decline but came about through a combination of factors not completely within Sinatra's control—his nemesis Mitch Miller's despotic reign as Columbia's artist-and-repertoire chief; the rumors about his ties with the Mafia; the jibes of superpatriot gossip columnists distrustful of his intuitive, for-the-underdog liberalism; the moral McCarthyism of the Catholic church directed against celebrities of the faith like Sinatra (and Ingrid Bergman, an adopted Catholic, for her roles as St. Joan and a nun opposite Bing) for the mortal sin of divorce. But at the time, it probably seemed that Sinatra was a victim of pop's planned obsolescence, even if no one would have put it quite that way back then, and even if it hardly made sense that just as the big bands were giving way to the singers, there was no room on the hit parade for the greatest boy singer of them all.

Lenny Bruce once decided that the survival of Florida's retirement communities was dubious because his grandmother was dead. By such solipsistic logic, it must have been tempting to conclude that Sinatra had outlived his welcome with the end of World War II. "The Voice is that of the teenage Romeo who, with opportunities unrationed, was thrust into the spotlight vacated in the suburban lanes and ballroom crannies by the eligible young man gone off to war—the neighborhood Tyrone Power or Robert Young who abdicated the civilized sex struggle at the behest of Uncle Sam," wrote Parker Tyler, arguably the most trenchant film critic of the forties, who obviously had no way of knowing that the scrawny Sinatra, 4-F personified, would eventually evict beefcake from the box office, just as he had from the jukebox.

No; if VJ Day signaled what looked like the beginning of the end for Sinatra, it was not for the reason that Tyler and others had envisioned. The problem was largely one of context. Wartime had fostered a morbid preoccupation with love's demise—a fatalistic understanding (perfectly if only tacitly expressed in those overwrought Axel Stordahl string arrangements with Frankie as the highest-pitch string of all—the one most likely to snap) that romantic bliss, ephemeral enough in the best of times, was now subject to termination by letters addressed "Dear John," and hand-delivered telegrams with a red star, beginning: "The Secretary of War regrets to inform you . . . ." After the war, with love once again a cause for foolhardy complacency, Sinatra and Stordahl must have seemed needlessly maudlin. Without the liner notes close at hand, I doubt I'd be able to listen to "There's No You" and "It Never Entered My Mind," both included on *The Voice*, and say which was recorded in 1943 and which in 1947. But that might be the point: record buyers of the time perceived a difference in climate, and climate makes all the difference in pop.

Fortunately, hindsight enables us to appreciate Sinatra's music from this difficult period of adjustment on its own terms, and the co-producers Joe McEwan and James Isaacs have made the job easier by eliminating the dead weight, a philosophy in blessed contrast to that espoused by the producers of other recent compilations (like *The Complete Sarah Vaughan on Mercury*, which included alternate takes as well as original masters of period effluvia like "Mary Contrary" and "Please, Mr. Brown"). Programmed thematically rather than chronologically, *The Voice* is subdivided into six albums: *Sinatra Saloon Songs, Sinatra Love Songs, Sinatra Standards, Sinatra Swings, Sinatra Screen, and Sinatra Stage*. Naturally, the categories overlap, and one could quibble that both *Saloon Songs* and *Love Songs* contain more "standards," in the generally accepted sense of the word, than the sides so designated, which are mostly given over to numbers crafted specifically for Sinatra and seldom performed by anyone else, like "Nancy (with the Laughing Face)" and "Saturday Night (Is the Loneliest Night of the Week)." But why quibble with music this seductive? The thematic unity the co-producers opted for is its own reward, particularly on *Saloon Songs*, which consciously alludes to the lovelorn drowse of such fifties Capitols as *In the Wee Small Hours of the Morning* and *Only the Lonely*. *The Voice* includes seventy-two tracks, about half of Sinatra's Columbia output, with very little drawn from after 1949. Roughly three quarters of the tracks are ballads like ballads ought to be, universal in sentiment but personal in the telling. The most celebrated of these plaints is "I'm a Fool to Want You," not so much a song as one sustained, manly erotic shiver. But there are numerous other performances in the same vein, many of them very nearly as affecting, notably "Autumn in New York," "There's No You," and "I Fall in Love Too Easily."

Quite aside from their musical virtues, these ballads demonstrate how pop values have changed over the decades. Sinatra was singing for the masses but giving each listener the impression that the message was for his or her ears alone; Bruce Springsteen's genius (like that of Elvis Presley, the Beatles, or Michael Jackson) is to make even the solitary listener feel connected to the cheering throng. I suspect that many listeners weaned on rock 'n' roll, having made a separate peace with Sinatra's crooning, will continue to find Stordahl's rhythm-abating string arrangements a painful reminder of pop's saccharine past. In defense of these ballads, one could argue that they come as close to art song as forties and fifties pop ever gets, and do so without sacrificing pop's intimacy or regular-Joe vitality. Still, entire album sides of weepers originally meant to savored on single-play 78s can quickly become too much of a good thing. Insolent swingers like "It All Depends on You," "For Every Man There's a Woman," and "Day by Day" are rewarding not only as welcome changes of pace but as occasions for Sinatra to forego the precise diction he didn't learn in Hoboken, to try out the contented tomcat purr that he displayed to full advantage only a decade or so later, on *Nice and Easy* and *Songs for Swinging Lovers*. "Bess, Where Is My Bess" and "Ol' Man River," Broadway show-stoppers from the *Stage* album, find him experimenting with a full-chested belting style not generally associated with him until the seventies. Recorded in 1943, during a musician's strike, "Close to You," "I Couldn't Sleep at All Last Night," and "Oh, What a Beautiful Morning"—three tracks with contrapuntal vocal choirs arranged by Alec Wilder—qualify as honorable attempts to make the best of a difficult situation. *Sinatra Swings*, the most uneven of the six volumes, is nonetheless a constant source of revelation because of George Sivaro's big-band arrangements and Sinatra's easy interplay with a host of jazz musicians, including Johnny Hodges, Coleman Hawkins, Marshall Royal, and a nonsinging Nat Cole. But Sinatra's kinship with jazz soloists is also evident on some tracks that swing lightly, if at all: Bobby Hackett supplies lovely trumpet interludes to "Body and Soul" and "I've Got a Crush on You," and Billy Butterfield does the same for "Nevertheless."

The Voice is such a pleasurable collection that it seems churlish to raise even a few minor criticisms. But I do almost wish there was a seventh volume called *Sinatra Schlock* with some of the feckless novelty tunes that Miller forced on Sinatra in the early fifties. After all, "Mama Won't Bark," featuring Dagmar and a dog imitator, may be infamous, but how many of us have actually heard it? Is it possible to appreciate the resurrection without witnessing the stations of the cross? A cover version of Johnny Hodges's "Castle Rock," with Harry James blowing through the ceiling and Sinatra disdainfully spitting the word "rock," is the only indignity included here; dire as it is, it's kinda fun, in the same silly way that Edmund O'Brien singing "Rockin' 'Round the Rockpile" in the movie *The Girl Can't Help It* is fun. (Ironically, Miller is present as an oboist on some of the tracks from the mid-forties; if only Sinatra had known what this Nero of the novelty number had in store for him!) The hand-tinting of the black-and-white cover photographs, though hardly comparable to Ted Turner's colorization of classic black-and-white films as an aesthetic crime, strikes me as a cheap, unnecessary touch. The liner commentary, by the music critics Gary Giddins and Stephen Holden, the film critic Andrew Sarris, the novelists Frank Conroy and Wilfred Sheed, the political columnist Murray Kempton, and the disc jockey and lounge lizard Jonathan Schwartz, leaves something to be desired, in that only the music critics address the music as such. The others, estimable personages all in their respective fields, are guilty of approaching Sinatra as a sociological phenomenon—though it must be admitted that Sinatra's career invites such cause-and-effect exegesis.

I ought to know. I can't listen to him without remembering an excerpt from Jean-Luc Godard's journals published some years ago in the short-lived English-language edition of *Cahiers du Cinéma*. The French director marveled that he could loathe everything John Wayne stood for as a man, "yet love him tenderly when abruptly he takes Natalie Wood into his arms in the next-to-last reel of *The Searchers*." Those of us who thrill to Sinatra's singing yet recoil from his lewd mysogyny and fat-cat Reagan Republicanism feel similar exasperation in trying to reconcile our disapproval of Sinatra's values with our advocacy of him as a performer. (Kitty Kelley, of course, spares herself such ambivalence in *His Way* by refusing to take the performer into consideration.) How ironic that long before he became Old Blue Eyes, Sinatra was dubbed The Voice, as though his vocal apparatus were a discrete entity with a life and will of its own. If only that were so, there would be no problem. The Voice didn't sensitize pop—the young Crosby had already done that (despite the readiness of some of *The Voice*'s annotators to write him off as innocuous old Uncle Bing). But by confiding between-the-lines longings and frustrations only female voices had hitherto been permitted to express so openly, the Voice introduced a new erotic dependency to courtly male pop.

Even now, though constricted by age and the abuse of nicotine and alcohol, The Voice still aches with every decent, tender emotion its owner's boorish behavior mocks. Yet Sinatra inhabits the songs he sings so forcefully and totally that it's impossible to embrace The Voice and spurn the man. He is the most calculated of singers, and part of what we respond to when listening to him is the consummate craft with which he has refined his vocal technique. Moreover, since Sinatra not only remolded the pop ballad in his own likeness but also redefined America's notions of what it means to be in love and what it means to be a man, our knowledge of his well-publicized gripes and grievances, his romantic conquests and setbacks, and his almost comical need for sycophantic male camaraderie supplies the context in which we listen to him, like it or not. (His staunchest middle-aged loyalists, like the nationally syndicated disc jockey Sid Mark, have the same problem I do in separating the man from his mishegoss, only they don't know it's a problem; it's Sinatra the ring-a-ding-dingster that they dig.) It's tempting to denounce Sinatra; to echo Kyle MacLachlan about Dennis Hopper (in *Blue Velvet*) and ask, "Why are there people like Frank?" But *The Voice* is a reminder of what attracted us to him in the first place, and a good enough reason to fall in love with him all over again.

(December, 1986)

TOO LATE BLUES

For a jazz critic to admit to a fondness for Bobby Darin isn't as embarrassing as it sounds. After all, Darin was a *down beat* cover boy in 1960, when Gene Lees, then the magazine's editor, delivered this encomium: "Darin today is unquestionably the only young male pop singer who handles standards with something approaching the polished intensity of Sinatra." Thirteen years later, when Darin died following open heart surgery, the same magazine ran a moving obituary by the jazz critic and record producer Michael Cuscuna. Gary Giddins once told me that as a kid in New York he begged his parents to take him to hear Darin at the Copa. So I'm in good company as a fan. But I can go my distinguished colleagues one better: I used to want to *be* Bobby Darin. I still remember my junior-high glee club director's delight when I volunteered to sing "Mack the Knife" at assembly. "Oh, *Threepenny Opera!*" she exclaimed. I stared at her blankly. What did I know from Kurt Weill? I knew "Mack" from AM radio, which was where I picked up the rest of my Darin repertoire: "Splish

Splash," "Queen of the Hop," "Artificial Flowers," "Clementine," "Beyond the Sea"—all his hits, in fact. I don't need to be told that I could have chosen a hipper role model. But even as a teenager, far less inhibited than I am now, I found it easier to snap my fingers and warn anyone who would listen to watch out for Miss Lolly Linner (well, that's how I heard it) than to swivel my hips and caution those who were looking for trouble that they had come to the right place. I was essentially a good kid, not a punk. Like most of us who grew up to become jazz critics, I respected my elders, and I think I sensed that Darin did, too, despite his flippant air.

Darin's Atco hits weren't included in the fourteen-record soul retrospective that Atlantic released at the end of 1985. And that was only right, because he epitomized chutzpah, not soul. Yet it's impossible (for me, anyway) to think of Atlantic without remembering Darin, who was the Atco subsidiary's biggest moneymaker between 1959 and 1961 (the period of Atlantic's greatest fiscal growth), and whose producer was none other than Ahmet Ertegun, Atlantic's head honcho. I recalled Darin even earlier last year, though, when RCA surprised us with *Sam Cooke Live at the Harlem Square Club*, the long-overdue antidote to *Sam Cooke Live at the Copa*. Comparing the posthumous concert to the one released during the great soul singer's lifetime, critics made much of the different racial make-up of the two audiences, which seemed to me only half the story. In the early sixties, before anyone in the record industry actually used the term to describe the leap from ghetto to suburb, "crossover" also meant bridging the abyss from teen to adult, from sock-hop insouciance to supper-club savoir faire. Most in the industry's upper echelons were convinced that teenagers would outgrow rock 'n' roll along with acne and wet dreams, and that rockers who wanted to last in show business would eventually have to change their tunes (which is one reason that Tin Pin Alley standards and Broadway showstoppers provided the padding on so many early rock albums). It's unlikely that Cooke's producers had to twist his arm to get him to sing for aging white Manhattanites at the Copa. Certainly, no one had to twist Darin's. Although he was Atlantic's lead entry in the Italian-American pinup sweepstakes (one of his early albums was titled *For Teenagers Only*), Darin clearly considered swooning adolescent girls a necessary evil. You need only skim the paeans from Walter Winchell, Earl Wilson, and Dorothy Kilgallen on the back of *Darin Live at the Copa* (originally released on Atco in 1961, and now reissued in a facsimile edition on Bainbridge BT-6220) to ascertain the other audience Darin and Ertegun were openly courting.

Remember the *Dick Van Dyke Show* episode in which a mobster shanghaied Alan Brady's staff writers and bullied them into donating routines for his nephew, an all-around mediocrity who sang, danced, and did impressions? On *Darin at the Copa*, Darin comes dangerously close to reminding you of that nephew: he mimics W. C. Fields, Dean Martin, Walter Brennan, Ray Charles, Jimmy Durante, and Señor Wences; he fawns over an eight-year-old girl at one of the tables as he tosses off "Dream Lover" with contemptuous dispatch; he hobnobs with Joey Ross (Officer Tootie of *Car 54, Where Are You?*) and other "celebrities"; he cheapens his own deft vibraphone solo by cracking, "Red Norvo, eat your heart out"; he introduces "Mack the Knife" as "an old Bavarian folk song"; and he refuses a request for "Splish Splash" on the grounds that "That's going back too far" (making it clear that it's regression, not nostalgia, he's avoiding). Whew! Listening to this jive, you begin to wonder what it was you ever heard in Darin, but then he reminds you. Some of his banter shows sharp wit, as when he spots the heavy-lidded singer Keely Smith in the audience and announces, "Time to wake up, Miss Smith," or when he satirizes his own unctuous delivery on an impromptu "That Old Black Magic." In pop terms, his "Mack the Knife" cuts Sting's, and even Armstrong's, though the version here is a pale hint of his studio hit. Best of all, there are four well-modulated numbers that suggest Darin's unfulfilled potential as a Sinatraesque barroom crooner: Rodgers and Hammerstein's "I Have Dreamed," Arthur Schwartz's "By Myself," and Cole Porter's "You'd Be So Nice to Come Home To" and "Love for Sale." You come away from *Darin at the Copa* marveling at what Darin was capable of when he respected his material, and angry that his vulgar energy was never channeled positively for very long.

At its worst, *Darin at the Copa* still offers insights into the bourgeois ideals of Entertainment (gaudy) and Taste (proprietary) in the years before the Beatles. (You can almost hear the audience thinking, "Are we having fun yet?") *Two of a Kind* (Atco 90484-4-1Y), an album of duets with the singer and songwriter Johnny Mercer released in 1961 to tighten Darin's grip on the adult market, doesn't even have that going for it. Think of the Mercer lyrics he and Darin *might* have sung—"Skylark," "One for My Baby," "Come Rain or Come Shine," "Something's Gotta Give"—and what novel interpretations the cheeky Darin might have given them. Instead, *Two of a Kind* is squandered on Mercer ephemerae like "If I Had My Druthers" (from *Li'l Abner*) and "Lonesome Polecat" (from *Seven Brides for Seven Brothers*), and non-Mercer inanities like "Who Takes Care of the Caretaker's Daughter?" and "Paddlin' Madeline Home." Billy May's Jimmie Lunceford–like (or Sy Oliver–like, to be more accurate) arrangements are the album's only strong point, and they become more impressive when you consider the poverty of what May was given to work with. Atlantic would have served Darin's memory better by returning to circulation *That's All*, the 1959 album of showtunes and standards that yielded "Beyond the Sea" and "Mack the Knife," arguably the two finest vehicles for his anachronistic instincts (both tracks are still available on Atco's *The Bobby Darin Story**).

Darin might have developed into one of the finest American singers; his beginnings were no more ignoble than Tony Bennett's, who was little more than a human vibrato at the beginning of his career. It's tempting but ultimately discouraging to speculate what he might be doing had he not died at thirty-five. I remember that his reemergence as a denim troubadour in the late sixties struck me as opportunistic, though not in the way you might think: it was as if by singing his original protest songs, like "Long Line Rider" (about convicts found buried in an Arkansas prison yard), he thought he could reconcile his need to stay contemporary with his desire to transcend pop. This was not a mode he fully developed, however, and chances are that he would now be on the oldies trail or knockin' 'em dead in Vegas (consider that Wayne Newton started off with "Danke Schoen," a "Mack" clone originally offered to Darin). Having briefly studied acting at Hunter College, Darin gave gritty performances in such bleak sixties melodramas as *Too Late Blues*, *Pressure Point*, and *Captain Newman, M.D.* (for which he received an Oscar nomination). Although he did his share of silly romances, in film, at least, he surpassed Elvis and might have caught up with Sinatra. But by the time of his death, he'd worn out his welcome in Hollywood. The filmography at the end of Al DiOrio's well-researched but dully written biography *Borrowed Time* (Running Press, 1981) lists numerous shelved or uncompleted projects, including *The Vendors*, which Darin wrote, produced, directed, and starred in. Not magnetic enough for leads and too streaky to be a reliable character type, he would probably be thankful for *Love Boat* guest shots at this point. Darin's exuberance was pitifully wasted during his youth, but the waste might have become more flagrant once he reached middle age. Do I still endorse him in light of the evidence I've brought against him? Oddly enough, the answer is yes, though I'd be as hard put to explain why now as I would have been at fourteen. Other than muttering

---

* As a substitute for *That's All*, I'd recommend *As Long as I'm Singin'* (Rare 'n' Darin 1), a 1987 compilation of TV airchecks and previously unissued masters. It includes confident performances of "Minnie the Moocher," "This Nearly Was Mine" and "Just in Time," as well as a live version of Darin's own "That's the Way Love Is" (a well-crafted tune originally released as the flip side of "Beyond the Sea" and included on *Love Swings*, Darin's stab at Sinatra's *Songs for Swinging Lovers*). But *As Long as I'm Singin*'s absolute highlight is a TV aircheck of "Mack the Knife," recorded before Darin grew tired of the song, when he was still having fun with it. On the out chorus, he stretches syllables and lags behind the beat with the freedom and sophistication of a born jazz singer.

how nostalgia ain't what it used to be, it's tough to mouth profundities when you're busy snapping your fingers.

(March, 1986)

BLACK LIKE HIM

With its Victorian banquettes and pastel murals of fat wood nymphs striking fey poses (by the forgotten pseudo-Frenchman Marcel Vertes), the Café Carlyle is a dowdy relic from a time when a hotel's amenities were understood to be for the local gentry rather than for conventioneers. The Carlyle's waiters, busmen, and wine stewards are like the supercilious house servants you see in old movies: they go about their business as though invisible, demanding in return the sort of peel-me-a-grape insouciance it takes generations of inherited wealth to pull off. Anything more than that is an affront.

But which is more humiliating: to think of yourself as rude for not acknowledging services rendered, or to be thought gauche by others for not recognizing when expressions of gratitude are inappropriate? My companion and I were trying to be invisible, too, but we were making ourselves conspicuous by thanking someone each time a dish or piece of silverware was brought or removed. Our parents taught us that good manners would make us appear gracious, but what did they know of such things? When you're around people who take service for granted, you realize that good manners are a dead giveaway of humble beginnings.

What were we doing at the Carlyle if we felt so out of place? We wanted to hear Bobby Short in his natural habitat, but (speaking for myself) Short was only the half of it. I grew up believing, as many working-class Irish Catholics do, in an inverted Protestant ethic: heaven was reserved for have-nots like for me and my family, because wealth was conclusive evidence of corruption somewhere along the line. (This is still the creed I live by, and it's a good one to cling to if you plan on being a jazz critic.) But encouraged by the drawing-room musicals of the thirties and forties that I watched on television, I also grew up equating adulthood with limos, penthouses, luxury liners, and sinfully expensive supperclubs like the Café Carlyle (it didn't occur to me till later that none of the adults I knew in real life enjoyed any of these plums). Was the Carlyle as I imagined it would be? Yes and no. Remember the scene in *Hannah and Her Sisters* when Woody Allen drags Dianne Wiest to the Café Carlyle? She does coke right there at their table as they listen to Bobby Short. "They wouldn't know the difference!" she protests when Allen scolds her. "They're embalmed!" Just my luck. I finally make it to the Carlyle and nobody who's anybody goes there anymore. The Bright Lights in the Big City are somewhere else now, panting for more lurid thrills.

The Carlyle is the sort of room in which the entertainment is generally as unobtrusive as the rest of the help. But Bobby Short, who's synonymous with the place, is the antithesis of such a whispering performer. Befitting his past as a child vaudevillian, Short is a belter who spends a surprising amount of time on his feet, away from the piano (which, given his sub–Errol Garner effusions, is just as well). He claps his hands together *hard* every so often, as though realizing that the room itself is his competition. His extravagance transforms a bandstand tucked away against a wall into center stage.

"The audience has so much to distract their attention," Short remarked during a brief conversation I had with him in his tastefully appointed Sutton Place apartment the afternoon following my visit to the Carlyle. "They have a drink or dinner or a cigaret in front of them, and someone very important to them, either romantically or in terms of business, across the table. The idea is to go out there and grab their attention as quickly as you can and maintain it until you go off. You see, I can't give the audience a chance to resume their conversation.

"Before my arrival, the Café was a place where not only guests and residents of the hotel went for dinner, but also people from the neighborhood. The girls from Finch College would have their beaus take them there, and dine there with their parents, who would stay at the hotel when they came to visit." With his pampered moonface and rounded, townhouse diction, Short is the only man I have ever heard use the word "beaus" without sounding sarcastic. No, he wasn't wearing a monogrammed smoking jacket, but his dark blue jumpsuit looked as elegant on him. As he spoke, he sat in front of open shelves of African art objects. From time to time, his Dalmation, resting glumly on a divan, attempted to jump on his lap. "Jealous thing," he scolded, rubbing the dog's neck. "That's what he is, you know.

"The Carlyle was, and still is, that kind of upper-crust, Upper East Side hotel," he continued. "The café had always held its own financially, but it had never been that important a part of New York social life. I'm sure that many of the people who frequented the Café were unhappy with the changes that my presence brought about. On the other hand, people from all over started coming and liked me, which is why I'm still welcome there."

Short has held forth at the Carlyle at least six months a year since 1968. His current twentieth-anniversary engagement follows on the heels of his best record ever—the record that, to be frank about it, made me take him seriously. *Guess Who's in Town* (Atlantic 81778-1) is an homage to the late Andy Razaf, a prolific black lyricist best remembered for his collaborations with Fats Waller, although (as the album demonstrates) he also wrote memorable songs with Eubie Blake, James P. and J. C. Johnson, and the white transplanted Englishman Paul Denniker, among others. *Guess Who's in Town* is unusual for Short in that it exposes rather than disguises his lineage from Ethel Waters and other black vaudevillians. It's his *Black Album*, as it were.

The album's four Waller tunes are an unexpected pleasure, in light of Short's admission (to Whitney Balliett in *The New Yorker*, almost twenty years ago) that he felt "inadequate" to sing Waller. "I was probably thinking of Waller's bubble, and how impossible it would be to capture," Short said, when I reminded him of the quote. "But I feel quite comfortable singing the ballads he wrote with Razaf, because when Waller chose to sing a serious song, he often had a difficult time overcoming his comic image." A case in point is "How Can You Face Me?" one scoundrel's admonishment to another in Waller's interpretation, but a wounded reverie of seduction and abandonment in Short's.

Razaf's witty but down-to-earth lyrics rescue Short from the chichi that is sometimes his downfall, and he returns the favor by washing away the implied blackface that too many contemporary singers of both races seem to feel is necessary to interpret vintage black pop. He gives these songs the same respect he would Gershwin or Porter, and they deserve it. Recognizing that Waller's "Black and Blue" is as much a protest against a black pecking order based on skin color as it is against white oppression (Spike Lee thinks he discovered something new?), Short finds a poetry in Razaf's lyric that even Louis Armstrong barely touched. ("It's right there in the verse," Short told me. "'Browns and yellows, lucky fellows, ladies seem to like them light.' It's especially touching when you remember it was originally written to be sung by a woman [in the 1929 revue *Hot Chocolates*]. Coal-black women have always had a hard time of it in black society.") The late Phil Moore's arrangements frequently put a good horn section—featuring the trumpeter Harry "Sweets" Edison, the trombonist Buster Cooper, the alto saxophonist Marshall Royal, and the baritone saxophonist Bill Green—to trival uses, most noticeably on a over-syncopated "Honeysuckle Rose," and an aimless pastiche of big band themes on Denniker's "Make Believe Ballroom." But Moore's setting for "Ain't Misbehavin'" is refreshingly modern, with moody counterpoint between Short's piano and John Collins's guitar in advance of the horns. There are sprightly arrangements of Denniker's "S'posin'" and Blake's "Tan Manhattan" (an ode to Harlem's cultural self-sufficiency, although uncharitable ears might hear it as just another darky song), and a stark reading of J. C. Johnson's "Lonesome Swallow," featuring only Short's voice and piano, in faithful evocation of the classic 1928 version by Waters and James P. Johnson. The only one of the album's eleven tracks that backfires is William Weldon's "I'm Gonna Move to the Outskirts of Town," because Short, with his dainty elocution, sounds ludicrous shouting the blues.

In American legend, café society has always been where the white and the black folk meet, but in reality, the black folk are usually there to do their jobs. In his twenty years at the Carlyle, Short has sung himself practically hoarse. His baritone is so split with phlegm that he sounds some of his low notes in two different keys. (Thankfully, *Guess Who's in Town* was recorded during one of his vacations, when he was well rested and in relatively strong voice.) But age has done Short an unintentional favor in slowing down his vibrato, and in a space as confined as the Carlyle, he has propinquity working in his favor: the lavaliere microphone he wears on a string necklace allows him to fill the room without sacrificing intimacy. His between-numbers anecdotes are charming and full of information about what year and movie or show each song is from. He lavishes attention on obscure introductory verses, not as a pedantic exercise, the way thirtysomething cabaret show-offs like Michael Feinstein and Andrea Marcovicci do, but as a way of injecting suspense into overfamiliar material (and, in some cases, as a way of replicating the easy transition from narrative to song that the verses originally provided on stage or screen).

Accompanied by the bassist Beverly Peer and the drummer Robbie Scott, Short did a dozen songs the night I heard him, including two that have become unaccountably obscure—Kurt Weill's and Ira Gershwin's "You've Only One Life to Live" (from *Lady in the Dark*) and Harold Arlen's and Leo Robin's "Hooray for Love" (from the 1948 film *Casbah*). He revitalized Cole Porter's "I Get a Kick Out of You," giving it an unexpected twist by performing it as a ballad, phrasing the "You obviously don't adore me" line with genuine poignance instead of the customary Ethel Merman pizzazz. And he sang "Bye Bye Blackbird" with such feeling and style that I no longer doubted the legend that Miles Davis decided to record his classic version of this tune as a direct result of hearing Short's. (Miles voted for Short in a 1956 "Musicians' Musician" poll that Leonard Feather conducted for *The Encyclopedia Yearbook of Jazz*.)

To my regret, the set included nothing from *Guess Who's in Town*. But toward the end of the show, Short sang "Princess Poupouly Has Lovely Papaya," a piece of fluff from the public domain. With his shoulders thrown back in a characteristic pose that reflected his pride in being a self-made man, he told the audience that this was a song their parents used to hear as they drifted from nightspot to nightspot in the naughty twenties. I don't think I'm attaching too much importance to the fact that he didn't say *our* parents. For me, the key to Short's appeal is in the delirium with which he delivers the opening lines of "Make Believe Ballroom": "Away we go, by radio, to realms of sweet delight. . . ." This is a song Short has lived. Now in his sixties, he grew up wanting in on the good life he heard about on the radio and in the movies. So did I, but white boys don't have to answer to anyone for dreaming beyond their means.

A shocking number of both blacks and whites find something incongruous about a black man in bibbed shirt and cummerbund singing Cole Porter and Noel Coward to white swells, as though getting funky was what a black man had to do to certify his blackness. (And as though the race was too impoverished to accommodate variety—too isolated from the white majority to share any of its values.) Short has his own kind of soul. The ninth of ten children born to a Danville, Illinois, domestic who considered the blues too vulgar to permit in her house (and who allowed her son to enter vaudeville only because the family needed the income during the Depression), Short reminds us that black America has its own traditions of gentility and aspiration, by no means limited to its bourgeoisie. "I am a Negro who has never lived in the South," he wrote in *Black and White Baby*, his 1971 memoir, ". . . nor was I ever trapped in an urban ghetto." He's not a native of the high society he's bought into, either. But neither were the upwardly striving small-town midwesterners and sons of Jewish immigrants (all of them black and white babies themselves, in terms of what they absorbed from jazz) who wrote the songs that have become smart-set anthems. The Republicans think that ours is a trickle-down economy, but the truth of the matter is that our culture trickles up. Like the money we slave for, these songs about the finer things in life are too valuable to be entrusted solely to the rich. Which is why it's comforting to know that the man currently doing the best job of singing them in a supperclub knows what it's like to sing for his supper.

(*May, 1988*)

SUNSHINE, TOO

Sheila Jordan was once the subject of an amusing exchange between D. Antoinette Handy and George Russell—a classic confrontation between an empiricist and an artist who makes his own truth. It happened during a 1984 colloquium on jazz at Temple University in Philadelphia. Arriving a day before he was scheduled to deliver a lecture on his Lydian Concept, Russell sat unrecognized in the audience for a panel on women in jazz that included Handy, the author of a 1983 book on the International Sweethearts of Rhythm. She talked about a work in progress: a biographical dictionary of black women instrumentalists.

"You failed to mention two very important women in your talk this afternoon," Russell told Handy when the discussion was opened to the floor.

Who were they, Handy wanted to know.

"Billie Holiday, for one," Russell answered.

Handy patiently explained that her book was to be about *instrumentalists*, and Russell mumbled something about Holiday having influenced more musicians than any of the instrumentalists whom Handy had mentioned.

The second woman, Handy asked, somewhat testily.

"Sheila Jordan," Russell replied—to embarrassed laughter from the audience, because, as Handy quickly pointed out, "Sheila Jordan is white."

Russell, who is black but has often been mistaken for white (including by Wilfrid Mellers in *Music in a New Found Land* and, I would conjecture, by many in the audience that day), grinned devilishly and paused before delivering the punch line: "Well, that depends on how you define it, doesn't it?"

More than Sheila Jordan was at issue here, of course—though Russell is himself an academic and (via his adoptive parents) a product of the black bourgeoisie, the combination antagonizes him when he recognizes it in anyone else. In calling on a merit-system logic similar to the one that Lenny Bruce used in declaring Lena Horne Jewish and Sophie Tucker a Gentile, Russell was redefining the question of "authenticity" that hangs in the air whenever distinctions are made between black and white jazz performers. To Russell's way of thinking, Jordan was black because she'd paid more than anyone's share of dues: as a child of poverty, an hereditary alcoholic, the divorced white mother of a racially mixed daughter, and, perhaps most of all, as a jazz singer committed to expressing herself honestly in a society that places little value on honesty or jazz.

"Now I understand it," Jordan laughed, when I told her the story. "I was working at One Step Down [in Washington, D.C.] about a year ago, and a very nice black woman told me that my singing just blew her away. She introduced herself as D. Antoinette Handy and asked me to serve with her on a panel for the NEA [the National Endowment for the Arts]. She said that something George Russell once said about me made her feel that she had to hear me. She told me that what George said about me was absolutely right. And you know, I wondered what he could have said."

She was born Sheila Dawson, "at home, on a Murphy bed in a furnished room, in Detroit in 1928." (She kept the name Jordan from her brief marriage to the pianist Duke Jordan, who was Charlie Parker's pianist when she met him in the late forties.) Her parents had married earlier that day, and didn't stay together long after her birth. She was raised by her maternal grandparents in Summerhill, Pennsylvania, "in the mountains, about a mile and a half from Erenfield, a mining town that we used to call 'Scoopytown,' which is where I think the actor Charles Bronson is from."

A short woman with dark bangs, attractive moon face, and a big, toothy smile, Jordan speaks with a rural accent, dropping many of her final g's and losing some of her vowels—when she sings Hoagy Carmichael's "Baltimore Oriole," or mentions the record company she almost signed with in the early sixties, "Mercury" becomes "Merkry." She talks about her life with the same candor and lack of self-dramatization that characterizes her singing. "People used to sing a lot where I was from, because it was a very depressed area, and there was nothing else to do except go to the beer gardens and drink and sing and forget your problems. I've been singing since I was three. I'd sing when I was happy, and I'd sing when I was unhappy, which was more often. I sang just for the joy of singing. I remember I used to have to pass a graveyard on my way to the store, and, boy, I would sing my lungs out because I was so scared. In Johnstown, they sold magazines with the lyrics to all the current songs, good songs that later became the standards that the jazz people took and made into classics. That's what I grew up singing. I always felt better when I sang. Singing made life bearable. I never sang in church, because I was Catholic and only boys were allowed in the choir. But do you want to know something? I just recorded a Gregorian chant, in Latin, with George Gruntz's big band. After all these years. But I was never allowed to sing in the Catholic church.

"On Saturday afternoons, I used to walk two miles to South Fork to the movies, when I could get the money and there was a Fred Astaire movie playing there. I loved his singing, the same love I later felt for Billie. He had such a personal sound. He and Ginger Rogers—that was glamor to me. It was an escape from the poverty and unhappiness at home. I grew up in what was basically an alcoholic family. My mother died from the disease. My grandfather was an alcoholic, and my grandmother also started drinking very heavily toward the end of her life. The bills would never be paid, and the lights were always being turned off, which meant no radio. We didn't even have a library in town. I wanted to play piano so badly, and my great aunt—my grandfather's sister—was teaching me. But she hated the fact that my grandfather drank, and they had a big fight about it once. That was the end of my piano lessons. In small towns like that, people hold grudges. They never make up.

"Since my mother was the oldest of nine children, my uncles and aunts were like my brothers and sisters. Before me and one of my uncles, who was ten years younger than me, nobody in my family went past the sixth grade. The boys quit school to work in the mines, and when the mines were on strike, they would go away to the CCC camps, where they would clear fields or pave roads and send money back home. When I was about fourteen, my mother came back to visit. She was still drinking heavily, which she did up until her death in 1981, but my grandfather had temporarily stopped, which, of course, led to a big fight with me as the pawn. My mother said, 'Well, I'm taking her back to Detroit with me.' And my grandparents said, 'Sure, now that we've raised her.'

"I was terribly lost in Detroit. Here I was with my mother, who was totally out on booze—a beautiful woman, very warm and sensitive, but with a horrible drinking problem. I loved my mother, but she didn't give me anything of herself when I was a child. She was just a kid herself, in many ways. No, we never really came to an understanding before she died. She knew that I loved her, and I knew that she loved me, but she was incapable of showing her love for me the way she should have because of her disease. I never really began to understand that until I got the disease myself, and I didn't *really* understand it until around four years ago when I started going to ACOA [Adult Children of Alcoholics].

"My mother had seven husbands altogether, and when I lived with her, she was married to a creep who was a professional card shark. This guy—I refuse to call him my stepfather—was beating my mother up, and I couldn't stand to watch it. She was a little ninety-pound woman, a very attractive woman who was unable to defend herself, and he was breaking her nose and knocking all of her teeth out. Plus, he was getting fresh with me, fondling me when he thought I was asleep. It was just too much for me. I moved out when I was seventeen, got a part-time job, and rented a room in the Evangeline Home, a place for young woman. That's where I was living when I finished high school. My classmates thought that was pretty cool, me being on my own like that."

Jordan is sometimes included in lists of illustrious jazz graduates from Cass Tech, the accelerated Detroit high school that produced Gerald Wilson, Milt Buckner, Paul Chambers, Tommy Flanagan, and Donald Byrd. "But I actually only went to Cass for a year, and not because I was so smart. It was just that they didn't know where to place me when I transferred from Pennsylvania. To be perfectly honest, I couldn't keep up with the work at Cass, so I transferred to Commerce, which was connected to Cass by a bridge. Commerce was where you went to learn typing and shorthand, and that was good for me because I knew I wasn't going to college and I was already thinking in terms of how I was going to support myself. The kids from Commerce ate lunch in the Cass Tech cafeteria, and took certain academic courses at Cass. It was basically one school, and we all hung out together."

She was exposed to jazz by her friends at Cass, and her love for the music was part of a larger infatuation with black culture in general. "What made me feel a bond with black people, first and foremost, was the music, which I think literally saved my life. But I remember something my grandfather once said. He was reading a story about black people in the newspaper, and he looked up and said, 'You know, those poor people get blamed for everything.' He wasn't an educated man. He was a house painter who had only gone to third grade, but he had beautiful handwriting, and he had compassion for people. I really identified with the black people I met in Detroit, maybe because I felt like I had experienced prejudice myself in Pennsylvania. There were two families that were poorer than all the others, and we were one of them. My grandfather was the town drunk, and he was considered an atheist because he wouldn't go to church. He was part Cherokee, and very much into his own way of believing. He couldn't understand why we would go to church on Sunday mornings and not eat breakfast, in preparation for receiving holy communion—I mean, not that there would have been much for breakfast anyway, but he would get mad that we couldn't eat a piece of toast or something. He wasn't a violent man. He never struck us, but we would have to sneak off to confession and communion when he wasn't looking.

"I was in an awful state in Detroit, and black people were the only ones who showed me compassion. I felt black. I wanted to be black. I used to tell people I *was* part black, because I hated the words I heard white people use to describe blacks, and I thought that maybe they wouldn't use that language around me if they thought I was black. My friends were Barry Harris, Tommy Flanagan, and Kenny Burrell, who were all around my age. The police were always harassing us. I'll never forget the last time I was taken down to the police station in

Detroit. When they stopped our car, my girlfriend Jenny was in the front seat with the fella she liked, and I was in the back seat with Frank Foster [the tenor saxophonist who now leads the Count Basie Orchestra], who I had a crush on. I threw my cigarette out the window, and, don't you know, they crawled down under the car to find the butt. They thought they had us for possession, but they took us in anyway and gave us the third degree. I told the detective I was going to be leaving town soon anyway, and he made fun of me. 'Oh, sure,' he said. 'To New York, I suppose, where everyone's so *cosmopolitan* and all the white girls have black boyfriends. Let me tell you something, young lady. I have a nine-year-old daughter, and if I thought I was ever going to catch her in the situation I just found you in tonight'—he touched his holster—'I'd go home right now and shoot her in her sleep.' He gave me the chills. I knew I had to leave Detroit as soon as I finished school."

Though she'd performed in Detroit with Skeeter Spight and Leroy Mitchell in a teenage vocal trio that improvised lyrics to bebop solos, Jordan bolted for New York with no intention of pursuing a singing career. "I was chasin' the Bird, just like everybody else in those days. I wanted to be able to hear Charlie Parker all the time, and sit in with him once in a while. The first time I met him was in Detroit, when my friends and I were trying to talk our way into the Club El Sino, where he was working. First of all, we were underage, and second, it was a black club and my girlfriend and I were white girls with two black guys. They turned us away, because they didn't want any trouble with the law. We went into the alley behind the club, and the door was open on account of the heat—they didn't have air-conditioning in those days. Bird came out and looked around. I wouldn't have thought of this then, but maybe he was coming into the alley to get high. I don't know. But he saw us and talked with us. Then when he came to town again, he let us sit in with him at the Greystone Ballroom—Skeeter, Mitch, and I.

"I took him to the Bluebird Inn, where he sat in with the house band. He drank Angel Tips—that's what he called it—creme de cocoa with cream on top. He had one and bought me one, but he never tried to come on to me. He was always a gentleman with me. He used to hang out at my loft in New York, and he turned me on to Bartók and the modern painters. He appreciated all kinds of music. He even liked Dinah Shore. He could hold an intelligent conversation on any subject. He could meet an engineer and rap about engineering with him. He was an amazing man, and that's why I'm almost afraid to see the movie about him. I feel too close."

In New York, Jordan studied briefly with Lennie Tristano, who had a reputation for insisting that his instrumental students sing. (The point was to make them realize that the music originated in them, not in their instruments.) How did he go about teaching an actual singer? "The same way. He asked if I could sing along with one of Bird's records, and I told him I already did that. He put one on, I sang along with it, and he said 'Ah, so you do.' Then he assigned me Lester Young. He also wanted me to sing along with Billie and Sarah, which I didn't want to do, great as I thought they were, because I wanted to sing my own feelings. So he didn't press it. He mostly coached me in rhythm and harmony.

"I studied with Lennie because I was looking for a place to sing, and I knew that he had sessions at his place every Friday and Saturday. I wanted to sing, but it was the same way it is to this day. My goals weren't the same as other singers'. I didn't have any aspirations of being a great jazz singer, and everybody knowin' about me and thinkin' I was wonderful. I've always sung for the sheer pleasure of it."

Though she rarely sang professionally during her early years in New York, she was always on the scene. Her loft was a hangout not only for Parker, but for countless other musicians and their camp followers. (In a *Cadence* interview, she reminisced about a time when she came home from work to find Sonny Rollins jamming in her loft; so many people were there that she couldn't squeeze in.) On the face of it, she was leading the bohemian life. But she had a daughter to provide for (Traci, from her marriage to Duke Jordan), so her situation was more complicated than it might have been for a male scene-maker. She worked as a typist and sang whenever an opportunity presented itself, even if the job paid next to nothing.

Fortunately for the rest of us, musicians appreciated her singing, and one in particular nudged her into the spotlight. In 1962, she was performing with a group that included the bassist Steve Swallow at the Page 3 in Greenwich Village. "Steve Swallow was very close to George Russell at the time," she recalled on a 1980 ECM promotional interview disc, "and George Russell came in one night to hear Swallow and to hear [the pianist] Jack Reilly [who was studying with Russell and told him] you should hear this singer. George came in, and he heard me, and he really liked me—in more ways than one, I might add. [*Laughter*] So George and I became very close, and he was a great supporter of my music. . . . If George Russell had not come into the club that night, I would never have been recorded. He took money out of his pocket and made a tape of myself and [the guitarist] Barry Galbraith, and took it to Al Lion and Mercury. Of course, I never thought anything was going to happen with that. I was just so happy there was going to be a chance for me to sing on record, whether it was going to come out or just be a tape. . . .

"Blue Note signed me immediately. . . . But before I did *Portrait of Sheila** . . . [Russell] wanted to know, 'Where do you come from to sing that way?' So I took him back to Pennsylvania and showed him the mines. A miner asked me to sing 'You Are My Sunshine.' I said, 'Oh, I don't sing that anymore.' He said, 'Well, you used to.' . . . My grandmother said, 'Well, let's all sing it.' So she played it on the piano, and we all sang it, and George got an idea, and that's how 'You Are My Sunshine' came about on *The Outer View*."**

The first time you hear it, Jordan's vocal on "You Are My Sunshine" is so electric that the rightness of all that surrounds it—Russell's probing piano lines (at once dissonant and folkloric); Swallow's combative, Mingus-like bass counterpoint; the quiet virtuosity of the brief horn solos (the tenor saxophonist Paul Plummer's, in particular); the sarcastic coda at double the tempo and the final negating three-horn raspberry—almost passes by unnoticed. But Jordan's vocal is what makes the track overpowering even after you know every note by heart. She enters *a capella* about five minutes in, singing the beery, unlovely melody as sweetly as a child intones a prayer—yet the words acquire an unaccountable erotic chill (her slight lisp, all the more noticeable with no instruments to mask it, works in her favor here, as I suspect Russell knew it would). After the horns reenter, a precipitous key change as the tempo quickens forces her above her natural range in a successful attempt to exploit her tendency to sing a quarter-note sharp. Although she makes the notes, it sounds like she's straining; but this helps in transforming the feckless song into a lover's needy plea and something else beside—something very close to what Michael Cimino failed to get at with the singing of "God Bless America" at the end of *The Deer Hunter* (which was set in the same part of Pennsylvania as Jordan's hometown). I once played "You Are My Sunshine" for a friend who wasn't really a jazz fan, and told him the story of how Russell and Jordan came to record it. He listened to it in silence, then said that he never before realized how symbolically charged sunshine must be to coal miners who spend their days working in darkness. Or, as Russell said about Jordan's performance in the original liner notes to *The Outer View*, "She's jazz, but she's *Sunshine*, too."

It's shocking to realize that after *Portrait of Sheila*, Jordan had to wait twelve years to make another record under her own name. It wasn't until she became

---

* Just reissued on compact disc only, Blue Note CDP-7-89002-2.

** "You Are My Sunshine," which Jordan recorded with Russell in 1963 (and is still available on *Outer Thoughts*, Milestone M-47027), wasn't her first recording. Two years earlier, she sang "Yesterdays" on *Looking Out* (Wave LP-1), an album by the bassist Peter Ind.

the surrogate "horn" in the pianist Steve Kuhn's quartet in 1980 that she performed with any regularity outside of New York. On the road with Kuhn, she was astonished to discover that she was something of an underground cult figure—the women singers of Jordan's post-bop generation whose values reflected those of jazz rather than show business could be counted on one hand (Jordan, Betty Carter, Abbey Lincoln, Jeanne Lee . . .), and of them, Jordan excited the most loyalty; in part because she had recorded so little that every note was precious. There were fans who had practically worn out their copies of her few albums, and who felt they had to tell her how much she meant to them. "Her treatment of a ballad is such that it makes me feel I'm intruding or that she knows what *my* story is," one fan wrote *Down Beat* in 1963, speaking for all of us.

Shortly before surfacing with Kuhn, she made a resolution to deal with the drinking problem she feels she inherited from her mother. "I've been sober for two and a half years now, although I stopped drinking around 1975," she explained. "I slipped only once, when I started drinking nonalcoholic beer, which led to actual beer drinking. That lasted for about a year before I stopped again in 1977. The reason I joined a self-help group wasn't because I was afraid I would start drinking again, but because I was getting scared of my dependency on coke. How it started was that around 1981 people who didn't know I had an addictive personality would offer me some now and then. I switched seats on the Titanic. I never reached the point where I needed it every day, but I was never an everyday drinker, for that matter. I would have total blackouts when I drank, so I drank only on the weekends. Never during the week, because if I touched a drop, I would be off and running. It was the first drink that got me drunk.

"It was the same way with the coke. With Charlie Parker, it was the opposite pattern. He drank more when he wasn't using dope. Once he came to a party in my loft and drank rubbing alcohol when there was nothing else left to drink. Another time, on a Sunday morning when the bars were closed, he went to the store and bought ten bottles of lemon extract, drank them, then went back and bought and drank I don't know how many more bottles of it. I can identify with what Bird was going through. I remember one time I dropped a little brown vial of coke on my bathroom floor. It splintered into pieces, and, don't you know, I picked out the glass I could see and snorted. I mean, how sick can you get? Your habit overpowers you. It tells you what you will do and what you will not do. It's a disease, and I could sit here all night and tell you I'm over it, then slip right back into it tomorrow morning. The worst part is that knowing, hey, this stuff can kill you isn't enough to make you stop. You don't care whether it kills you because you have a death wish anyway."

Even as the pace of Jordan's career quickened, she never considered giving up her nine-to-five job that she held since 1966 as a typist for Doyle Dane Bernbach, a major New York advertising company. "I was poor as a child, and I swore to myself that I would never, so long as I could help it, live like that again. I've always had a job. But you have to be careful of what you pray for, because you might get it. Every morning, I used to pray that the day would come when I could go on Social Security and devote full time to music. But last year, my company merged with another company, and they laid me off. After twenty-one years! I was devastated at first, but I guess it was the push I needed. I'm a full-time singer now, and so far it's worked out fine."

At sixty, Jordan finds herself in steady demand. When she sat down to talk, she had just returned from three months in Europe with the George Gruntz Concert Jazz Band. A few days later, she was off again, for a series of voice and bass duet concerts with Harvie Swartz. She also travels as a single, performing with local rhythm sections in each town, and she devotes much of her time to teaching (she has conducted vocal workshops in Gratz, Austria, and at City University in New York). When she was still married to Duke Jordan, "He used to ask me, 'Why can't you sing the melody just the way it is, *then* do what you feel you have to do to it after the first chorus?'" Though she declines to admit that this was good advice, she does admit offering a variation of it to her own students. "I tell them you want to approach a song melody-first, the way it's written. If I feel a kid isn't being honest with a song, I'll slow her down. I'll say, 'Sing me the song first. Until you're able to sing a song with feeling all the way through, you're not ready to scat.'"

With her severance pay from Doyle Dane Bernbach, she made a down payment on a house in Hunter's Land, New York, in the Adirondacks, where she now spends most of her free time. "I was seeing a man from there, and after we broke up, I realized how much I liked the serenity up there. I grew up in the mountains, and I never thought I'd want to live in the mountains again. But I guess that as you get older, you find yourself getting in touch with your roots. There's nothing like having someone you love and respect who loves and respects you back. But if you can't have that, you're better off alone, like I am. I don't like being alone, but I'm in no position to have a relationship. I'm away too much, and I had one that lasted sixteen years and almost did me in when it was over. I don't want to go through that again so soon.

"At this point, my daughter is a vice-president at Motown Records. She was in promotion at Arista, working on behalf of Aretha Franklin and Billy Ocean, but she was getting so many offers from other companies that she knew it was time to move on. The only bad part of it, for me, is that she had to move out to Los Angeles, so I no longer see her as much. But the title means a lot to her as a black woman. She didn't have it easy. You can imagine how proud I am of her. I'll tell you a story. Once when she was a baby, I was taking her to Coney Island, because she loved the water. A woman on the subway with two infants of her own said, 'Oh, what an adorable colored baby. Where did you get her?' I said, 'I got her the same way you got yours, lady.' The prejudice never seemed to bother Traci, but it occasionally bothered me."

Certain songs are now associated with Sheila Jordan. *Body and Soul,* a recent Japanese release (CBS/Sony 32DP 687) was practically a Sheila Jordan retrospective, with new versions of "Baltimore Oriole," "I'm a Fool to Want You," "Falling in Love with Love," and "When the World Was Young" from *Portrait of Sheila*. She has written others that deal frankly with her life (for instance, the title track on *The Crossing,* Black Hawk BKH 50501-1-D). But if her singing has become more autobiographical, it has also become, paradoxically, less direct. It's no longer so intense, which is understandable; one suspects that singing no longer amounts to an emotional release for her, as it did when she had fewer opportunities to perform. For my taste, she now scats too much for someone with her gift for interpreting lyrics (though, in all fairness, it should be added that her scatting rivals Betty Carter's for harmonic aptitude and rhythmic acuity).

Still, at her best, there's no one else like her. Her personal integrity has made her a role model for young women singers, including some who owe nothing to her stylistically. In a 1981 *down beat* Blindfold Test, Jordan said about a record by Billie Holiday: "Billie's sound is so strong that you forget who's playing behind her. . . . All you're aware of is this wonderfully human voice with all this emotion and love in it, and pain—the whole thing, it's just right there. I mean she could sing with anybody—she could sing alone, she could sing with the guy on the street corner, she could sing with anybody and make it jazz. She *is* jazz, she still is, she's *the* jazz singer to me." When told that a younger singer had recently said much the same about her, Jordan said, "Oh, wasn't that sweet of her. I had no idea she felt that way about me. It's always wonderful when someone tells you how much they enjoy your work. Sometimes they don't, because they assume you hear it all the time. But I *don't* hear it all the time. What I sometimes hear from black singers, who aren't really putting me down or anything, you understand, is, 'Well, it isn't really Sheila's culture'— whatever that's supposed to mean. Maybe it wasn't my culture to begin with. But I didn't steal from anybody. The only person I ever wanted to steal from was Charlie Parker, and I didn't have to. He gave it to me.

"I know how hard it is for women singers, and I try to give them advice if I can. First of all, if they're good lookin', they're going to have to go through a lot of B.S. with club owners and record producers. And any time anything goes wrong, the singer's always the one who gets blamed. I don't feel I'm an entertainer, but from having been with Steve Kuhn's band and doing the bass-and-voice duets with Harvie, I've learned to present my music in a way that allows the audience to become involved. Years ago when I sang, I was closed up, standing very stiff with my head down. Steve Kuhn used to say, jokingly, that when people came to hear the quartet, they left raving about me, not the band. The focus was always on me, as hard as we tried to present that group as a quartet, not just a singer with a trio. But I said, 'You gotta remember one thing, Steve. I'm the one they talk bad about, too, when they don't like us.'"

(November, 1988)

SHE'S SUSANNAH McCORKLE

Susannah McCorkle has chosen two antithetical careers for herself, one begging for solitude, the other demanding the spotlight. "I have two literary agents, one here and another in England, telling me not to equate my ability to write with being unfulfilled," she says. "Before I got involved in a good relationship, I thought of myself as a solitary woman, speaking for other women who thought of themselves the same way. But who am I now? And who am I speaking for?" She's published several short stories, one of which—"Ramona by the Sea," about a moody, overweight young woman's alienation from both the middle-class values of her parents and the radical alternative of her college classmates—was included in the 1975 O. Henry Prize collection, in the fast company of works by Alice Adams, Harold Brodkey, Raymond Carver, E. L. Doctorow, and Cynthia Ozick. There is an unfinished novel in her drawer, which she plans to resume work on after recovering from what she hopes is a temporary case of writer's block—brought on, in part, by changes for the better in her love life and success in her second career as a jazz-oriented singer.

Because McCorkle—who is guarded about her age, but is probably in her early forties—performs standards, rarely scats, and seldom uses her voice as a horn, there are some who dismiss what she does as gentrified pop. Her failure to record with pacesetting black instrumentalists has hurt her credibility with younger jazz critics, who not having much use for women singers anyway, tend to rate them by the company they keep. (How else to explain the reams of praise for such earnest but unconvincing young black singers as Carmen Lundy and Cassandra Wilson?) But what distinguishes McCorkle as the outstanding female jazz singer of her generation is her unforced affinity for the era of Billie Holiday, Lee Wiley, and Mildred Bailey, when jazz and pop shared a repertoire, and the test of vocal improvisation was straightforward melodic embellishment and scrupulous attention to lyrics. McCorkle recognizes that the best older songs still have plenty of life in them: she feels no need to treat them as faded art songs, as too many modern cabaret performers do. Nor does she accessorize songs, like Linda Ronstadt and other rock singers who embrace vintage pop as an excuse for period extravaganza. She merely *sings*—in a merry, slightly throaty voice that puts itself at the service of a song, instead of the other way around.

She's had bad luck with record companies, twice signing with financially insolvent independents that hoped to make big bucks with fusion and had little notion of how to promote a quality singer. But this hasn't prevented her from recording the best composer/songbook series since Ella Fitzgerald's. McCorkle's best albums are *The Songs of Johnny Mercer* (Inner City IC-1101), *Over the Rainbow: The Songs of E. Y. "Yip" Harburg* (Inner City IC-1131), *The Music of Harry Warren* (Inner City IC-1141), and *Thanks for the Memory: The Songs of Leo Robin* (Pausa PR-7175). On none of these does she settle for just the songs that everyone remembers. She includes Mercer's "Harlem Butterfly" along with his lyric for "Blues in the Night," Harburg's lyric for "The Begat" from *Finian's Rainbow* as well as that for "Over the Rainbow," Warren's "The Girlfriend of the Whirling Dervish" along with his "Forty-second Street," and Robin's lyric for "Hooray for Love," as well as that for "Thanks for the Memory."

But McCorkle, who is genuinely interested in songs that reflect contemporary mores, and who fears being stereotyped as an antiquarian, has also recorded three albums—*The People That You Never Get to Love* (Inner City IC-1151), *How Do You Keep the Music Playing?* (Pausa PR-7195), and *Dream* (Pausa PR-7208)—which have juxtaposed Rupert Holmes with Jerome Kern, Jimmy Webb with Irving Berlin, Paul Simon and Antonio Carlos Jobim with Cole Porter and Rodgers and Hart. "People have told me I would have been a big star if I had come along back in the thirties or forties," she says, "but I've never wished for that, because I prefer being a woman now. I would have been a band singer or a starlet—a blonde cutie. I'm not one of those people who thinks that every song from the thirties and forties is wonderful and we should bring back the big bands. I just like good songs, and I can hear right through the period trappings, including the instrumental trappings of the seventies and eighties. Recently, a friend of mine in his late thirties made a tape for me with lots of what I suppose would be called country-rock songs on it. The melodies bore me, but I can relate to some of the lyrics, and there's one song by John Hiatt that I'm thinking of doing. I don't enjoy his singing, but his lyrics are about universal themes: failing relationships, marital disappointment, coming to terms with the fact that you can be just like your parents, no matter how differently you've tried to live your life. The song I like is called 'You May Already Be a Winner,' inspired by those millionaire sweepstakes offers we all get in the mail. It's really a song about taking stock of a relationship and deciding that maybe life isn't so bad. We may already be winners, because we love each other. It's a good example of a song that couldn't have been written in another decade."

McCorkle's list of favorite singers—Billie Holiday is one, as you would expect, but so are Ray Charles, Mose Allison, and early Nat Cole—is surprising for someone whose own style is about effervescence, not grit. However much her phrasing owes to Holiday (and however much she sometimes sounds like Marilyn Monroe might have in clothing that actually fit), McCorkle's precise diction and fresh-scrubbed timbre recall nobody so much as Doris Day—an observation she accepts gracefully enough now, although it used to vex her no end.

"The owner of a gay club I worked at in London told me not to come back without spiked heels, false eyelashes, and tits out to here. He fired me for sounding too much like the girl next door. 'You sound like Doris Day!' That hurt. As I get older, my voice is becoming deeper and darker, but it's still very sweet and youthful, and I've gone through periods of hating it, when I felt pretty battered by life and wondered why my voice didn't reflect that. I'm flattered now when someone says I sound like Doris, because I've come to realize what a wonderful singer she is. But she's still not one of my favorites, because she's not the kind of singer you hear and think, 'Who is this woman and what has she lived through?' like you do with Billie Holiday. You never wonder that about the well-brought-up white ladies. The first time I heard Billie, I felt as though there was nothing separating us. She was right there. You heard the human being, not the performer. I want people to feel that way about me."

In March of 1988, McCorkle taped a live concert for WMHT, Albany's PBS station, at Proctor's Theater in nearby Schenectady, the city she commutes to from New York to be with her fiancé, a reporter for Albany's CBS-TV affiliate. Called *Susannah McCorkle and Friends: Jazz Meets Pop*, the concert also featured Gerry Mulligan, Mark Murphy, the Gene Bertoncini–Michael Moore Duo, and the Eastman Jazz Ensemble.

Schenectady's evacuated factories, streets, and stores depressed me. So did the people I saw aimlessly wandering the streets—it was difficult to tell who was homeless and who was just on the way there. General Electric, which boasts of bringing good things to life, sentenced this once-thriving GE company town to death. Downtown looks like all cities will soon look in an America without the prosperity that once entitled anyone who could hold a job to think of himself as upwardly mobile. From my hotel, I could see the skyline, which was nothing but crosses and spires: there's a lot to pray for in Schenectady.

The only visual relief from this urban disintegration was Proctor's Theater, a Art Deco movie palace with quaint little shops in its arcade lobby, vaulted ceilings, ornate balconies and mezzanines, a proscenium stage, gaslights, and the pipes of a 1931 Golub Mighty Wurlitzer Organ up two walls. If Schenectady's economy were better, Proctor's would've been sold to a movie chain and multiplexed years ago; restored in 1979 to something approaching its original grandeur, it's an Edenic shrine. ("One summer when I was eleven or twelve, I hitchhiked [from Amsterdam, New York] to . . . the metropolis of Schenectady," writes Kirk Douglas in his 1988 autobiography, *The Ragman's Son*. "How large it was! The streets were so wide! And Proctor's movie theater was so much bigger than anything in our town. . . .") I also liked the people I met in Schenectady, especially McCorkle's beau, Dan DiNicola, a Chet Baker fan and frustrated trumpeter who hovered around Proctor's during rehearsal, playing stage lover—attending to the dozens of technical details every performer needs someone to attend to (especially a female performer, who risks being thought a bitch if *she* attends to them herself), and clearly taking vicarious pleasure in watching McCorkle on stage. (Ironically, as a result of being on TV, he's better known around town than she is.)

On stage at Proctor's, McCorkle cut such a trim, confident figure in a neat pageboy and a succession of sequined gowns and belted pantsuits that it was difficult to believe she ever had anything in common with the protagonist of "Ramona by the Sea," whose obesity was symptomatic of a chronic lack of self-worth. But like the unkempt Ramona, McCorkle was a compulsive overeater as a teenager. "I'd be in the supermarket at midnight, eating cakes and whole loaves of bread while I was still in the checkout line. It was frightening, because this was before anyone knew very much about feminine eating disorders, and I thought I must be the strangest person in the world. It's like being a drug addict or an alcoholic, except that you can't say you're going to cut out food entirely, the way you might with drugs or alcohol."

Also like Ramona, McCorkle was estranged from her college peers. Although active in the Free Speech Movement at the University of California in her native Berkeley, she was uneasy with the liberated cultural mind-set that went with political radicalism on college campuses in the 1960s. "My parents were active in leftist causes before me. My father is an anthropologist, but he worked on the waterfront during the war. I grew up with no religious training at all, knowing that you didn't cross a picket line no matter what, and believing that this country rightly belonged to Native Americans. I was raised, I would say, as a socialist agnostic. So political activism didn't represent the big breakout for me that it did for some other students. It was a wonderful time to be in college, because we really did feel we were effecting change. We were the first generation of college students to take on adult responsibilities, instead of acting like postgraduate high school students. But it was also very frightening to be twenty years old and see the police swarming all over campus, and hear the people you knew from classes talking about blowing up banks. A friend of mine from college recently asked me, 'How did you live through those years?' I never went in for floral shirts, love beads, acid rock, sleeping around, or naming children after continents. I hated going to parties where everybody smoked dope and lolled around, not talking. I wanted to go to parties to dance, meet a boy, and fall in love. In my writing classes, I was laughed at for writing stories that everyone said belonged in women's magazines. Everybody else was writing about first sexual experiences or first acid trips, and I couldn't, because I hadn't had either."

After graduation, McCorkle worked as an interpreter and translator in Mexico, France, and Italy, and it was while living the expatriate's life that she discovered jazz. "I went to Europe in pieces, determined never to return to the U.S. I can't say that I reinvented myself while I was abroad, because I didn't even feel that I had been anyone yet. But in Europe, I found a vindication for those instincts that had been completely thwarted in college: my romanticism, my love of quality writing and beautiful melodies, as opposed to loud, screaming guitars. So what if my stories were about housewives, so long as they were thoughtful, well-written stories? The things I responded to were valued in Europe, where you could buy a book of George Gershwin songs translated into Italian, see all the classic American movies of the thirties and forties for a few francs in Paris, and buy a Billie Holiday album for what amounted to a few dollars in a drugstore or supermarket. When I discovered jazz, it felt like being reunited with a family I never knew I had. I was glad, in a way, that I hadn't grown up listening to this music, because it might not have hit me so hard when I finally discovered it—I might have taken it for granted. Musicians I met lent me records by Kay Thompson, Duke Ellington, Stan Getz. It was like being accepted into a tribe. Once when I was flying from Rome to California, I had a few hours stopover in Mexico City, which I spent in a nice little café I remembered from when I had lived there for a few months. I struck up a conversation with a man who was excited to find out that I sang, and who told me that he was married to Jack Teagarden's daughter. It was an unhappy marriage, he said, 'but when you love jazz as much as I do, and you meet Jack Teagarden's daughter, of course you marry her.' Sometimes I think Dan wishes I was Chet Baker's daughter."

McCorkle made the leap from listener to performer "one, two, three, which says something about the galvanizing power of the music, because offstage I was still a very gloomy girl. As a teenager, I had sung in school plays and summer theater productions. My mother had encouraged me to try for a career on Broadway, but I didn't want to be Rose in *Bye Bye Birdie*, singing the same songs the same way every night, in the same costume, with my hair sprayed black. I didn't like singing in character. I just liked opening my mouth and singing. In Mexico City, after sitting in my room all day writing stories, I used to take walks down a traffic island during rush hour, singing where no one could hear me. It was a release from the solitude of writing, and I guess that's also how I became a performer. Getting an immediate response from an audience was very gratifying for someone like me who was used to writing a story over a period of weeks, having it published months or sometimes even years later, and maybe receiving one letter that the magazine would forward to me."

Because she was already writing prose, McCorkle tried writing her own material when she first began performing. "But it seemed as though every time I got a good idea for a lyric, I found an old song that expressed the sentiment much more eloquently. And I came across lyrics that Truman Capote and Dorothy Parker had written that were just okay—good, but nothing special. So I quickly realized that fiction and songwriting were different crafts. The older songs I liked had beautiful melodies like I knew I would never be able to write, and I also doubted that I could team up with anyone else my own age who could. The songwriters I knew were into rock, folk, country, and the blues. They weren't interested in writing pretty melodies, and even if they had been, could they have matched Harold Arlen or Cole Porter?"

While living in Italy, McCorkle decided that the time had come for formal training. Out of the blue, she phoned the director of an Italian opera company, and asked if he could recommend a good vocal coach. "He volunteered to coach me himself, and I didn't have the nerve to tell him that I wasn't planning to sing opera. He was short, fat, and terribly forbidding—an almost cartoonlike maestro waving his arms around in a dusty old apartment full of photographs of divas kissing him on the cheek. He kept whacking my jaw in an effort to get me to relax it, and made me buy a surgical corset because he thought I wasn't holding my stomach properly. He was very fond of me because I spoke fluent Italian and was very, very earnest. But after a few lessons, he finally told me—with regret in his voice—'I have to tell you, you will never be an opera singer. You have a tiny, tiny voice.' I told him that wasn't what I wanted anyway, and sang 'Hi Lili, Hi Lo' for him, as an example of the music I did want to sing. I chose it because it's a song about being sad, and I thought I'd be able to draw on the sadness I felt at that moment, on account of his having made me almost too self-conscious to sing. He listened to me and said, very contemptuously, 'For that kind of song, you have enough voice already.' I felt rejected, but also very relieved.

"I've never taken another lesson, although I do sing three octaves of chromatic scales every day, and try to keep in good physical shape by racewalking, exercising, and eating well. I don't read music, but I have ideas about how music should sound. I come to rehearsals with pages of notes about where I want key changes and rhythmic breaks. The musicians I work with respect that. They sometimes make fun of my imagery—when I tell them that I want a hazy, druggy feel on 'Old Devil Moon,' for example, or that I want to hear the bass sound like a heartbeat on 'If I Only Had a Heart.' But once they know the feeling I want, they have the freedom to translate it into their own terms. I'll never forget something Peter Ind, the American bass player, told me when I sang with his band in England, and confessed my feelings of inadequacy to him. He said, 'If you can let go and sing in a way that pulls the band and the audience into a song, do you really think anybody's going to care that you don't read music?'"

McCorkle first attracted attention as a singer during the five years she lived in London, before moving back to the U.S. in 1979 with her then-husband, the British pianist Keith Ingham. The marriage dissolved soon afterward, although the couple continued to work together in New York until 1983. "I treasure the experience of singing with Keith, but we were incompatible except musically—and even music became a problem. Keith loves Chicago jazz of the 1930s, and wanted me to be the reincarnation of Lee Wiley or Mildred Bailey, his favorite singers. I like them, too, but I wanted to expand. We would have big arguments whenever I wanted to do a contemporary song. The first time I showed him [Rupert Holmes's] 'The People That You Never Get to Love,' he crumpled up the sheet music and threw it across the room. He didn't even want to do 'There's No Business Like Show Business,' although I bet he would have if he'd known that Sonny Rollins had recorded it. He also didn't like Brazilian music, which I love, and he scoffed at the notion of having an act. A singer has to think about pacing, beginning and ending strong. You don't want to be Las Vegas, but you do have to be aware you're doing a show. I think that if Keith had enough money, he wouldn't even perform; he buy a magnificent piano and sit at home and play it. He defined himself as a vessel through which music passes—a very European attitude, and a good attitude to have for someone recreating classic jazz. Mine was the American attitude of 'I'm me, the performer that I am, projected through music.'"

Truth be told, I tend to side with McCorkle's ex in thinking that the contemporary tunes in her repertoire are unworthy of her. It's not that today's songs are intrinsically unworthy (though when we start talking Rupert Holmes, they are), just that the most memorable of them aren't songs *per se*, but hit records so identified with their original performers (who, in most cases, are also the composers) that they resist fresh interpretation. Which is why at Proctor's it was Paul Simon's "Still Crazy After All These Years," not George and Ira Gershwin's "'S Wonderful," that called up memories of a specific time and place. Something else was troubling. I once heard McCorkle dedicate a song to Fred Astaire, whom she said she admired "for just being a guy singing"; that's also the kind of integrity she usually projects, standing perfectly still except to rise ever so slightly to her toes for high notes. But smiling for the cameras at Proctor's, she indulged in arm-waving stagecraft and between-numbers patter that sounded over-rehearsed. (It was, at a six-hour dress rehearsal earlier that day; I thought it was amusing that Mark Murphy, a singer whose fans adore him for his alleged spontaneity, but whom I find unbearably "jazzy," did his numbers the same way every time. In the language of boxing, "meets" means "versus," and any time that *Jazz Meets Pop* on television, the smart money is never on jazz to win.)

But give McCorkle the benefit of the doubt. For her, singing Rupert Holmes and Paul Simon is an honest effort to reestablish contact with her own generation, not a desperate attempt to stay up to date, as it would be for a singer ten or twenty years older. She might not write her own songs, but she wants some of those she performs to reflect her own experiences and those of her generation, which is perfectly understandable, even commendable. She's internalized the dilemma confronting jazz in the face of rock hegemony. Besides, at Proctor's, she managed to make the tuneless "The People That You Never Get to Love" sound like the keening blues it secretly longs to be. And despite the uneasy mix of standards and what might (not uncharitably) be called substandards, her effortless version of "On the Sunny Side of the Street," with Gerry Mulligan's crooning obbligato, was a vivid illustration of the twofold pleasure McCorkle offers at her best: incomparable songs from an earlier day, and someone incomparable to sing them for us now.

(*July, 1988*)

POSTSCRIPT

I failed to mention above *As Time Goes By* (CBS Sony 28AP 3315), a Japanese release that McCorkle doesn't particularly like, although I do. The producers didn't allow her to choose her own material or sidemen, and I agree with her that the songs are overfamiliar and that the accompaniment is inattentive (the tenor saxophonist Jimmy Heath excluded). Still, McCorkle sounds wonderful, especially on a version of "September in the Rain" on which she lags indolently behind the beat, in no hurry to catch up.

In my liner notes for *No More Blues* (Concord Jazz CJ-), which was released in March 1989, I diplomatically refrained from mentioning my dislike for Gerry Mulligan's "The Ballad of Pearly Sue," a pseudofeminist, that-I-can-do lyric set to a businessman's boogaloo. I should also have mentioned my misgivings about the growl McCorkle affects on "Do Nothing Till You Hear from Me." But the rest of *No More Blues* is superb. McCorkle's intuitive musicianship is evident not only in her subtle and appropriate ballad embel-lishments ("Everything's Been Done Before," a duet with the pianist Dave Frishberg, is a good example), but also in her open-vowel riffing on "Breezin' Along with the Breeze"; her teasingly breathy responses to Terry Clarke's drums on "Fascinatin' Rhythm"; and the series of moaning, elongated notes with which she climaxes "Don't Let the Sun Catch You Cryin'," a near-blues associated with Louis Jordan and Ray Charles, not to be confused with the 1965 hit by Gerry and the Pacemakers of almost the same name. (Nor should Gordon Jenkins and Johnny Mercer's "P.S. I Love You," a hit for Rudy Vallee in the thirties, and for the Hilltoppers in the early fifties, be confused with the Beatles song.) "Sometimes I'm Happy" is a double homage to Lester Young and King Pleasure; McCorkle does a better job of adhering to Young's sinuous lines than Pleasure did, thanks to her surer intonation. *No More Blues* (CJ-370) has an advantage over McCorkle's earlier albums in presenting her in a fully realized band setting, rather than as a singer with instrumental backup. The music director is the alto saxophonist and clarinetist Ken Peplowski, but the real hero of the date is Frishberg, whose success as a singer and songwriter has taken him from his true calling as a vocal accompanist (remember his work on Jimmy Rushing's *The You and Me That Used to Be*?). Saving the best for last,

McCorkle conveys exactly the right combination of propriety, desire, and trust on "P.S. I Love You." A string of homey non sequiturs if taken at face value, Mercer's lyrics actually amount to a declaration of fidelity that asks the same in return. McCorkle's is the definitive interpretation of this unjustly neglected song; it takes a singer who pays so much attention to every line of the lyric to make you realize there's so much going on between the lines.

(*March, 1989*)

ALMOST WILD
ABOUT HARRY

"I'm twenty-one years old. I don't think of myself as a revivalist, and I don't want to be thought of as an anachronism because I like old songs," Harry Connick, Jr., the *wunderkind* pianist and singer currently in residence at the Algonquin, complained during a recent chat. Growing animated, he removed his rimless glasses, and continued in his New Orleans drawl. "Today, groove is all that counts, but they used to write melodies that were sophisticated and complex, lyrics that were tasteful, witty—not about drugs or sexual promiscuity. 'Baby, baby, yeah, yeah, yeah.' That ain't lyrics. But I don't want to do a song by George and Ira Gershwin, and sing it exactly as it would've been sung when they wrote it. If I were talkin' to a girl, I might say, matter-of-factly, 'The *way* you wear your hat, the *way* you sip your tea, the *memory* of all that, they can't *take* that away from me.' You know the Harold Arlen song, 'Let's Fall in Love'? If you look at it in the Harold Arlen Songbook, the rhythm is, like, '*Let's* fall in love, *why* shouldn't we fall in love?' But I would say, in conversation, 'Let's *fall* in love, why *shouldn't* we fall in love?' So when I sing it. . . ." Snapping a finger, he sang the lyric with the same inflections with which he recited it. "I *speak* a song before I sing it. I say the lyrics to break the rhythm up. I want people to relate to the words—to realize they're as meaningful now as they were fifty years ago."

Connick sang quite a bit during our interview a few hours before he opened at the Algonquin. He proved himself to be an excellent mimic. Between Marlboros and sips of coke with lemon, he did Louis Armstrong, Sylvia Syms, Stevie Wonder, and the melismatic singers who oversell both their songs and themselves on the weekly cable program *It's Showtime at the Apollo,* Connick's favorite television show from the warped perspective by which some of us enjoy Geraldo, Oprah, Sally Jessy, and Phil. This ability to conjure other singers is frequently a curse as well as a blessing, Connick admitted. "If I listen to Sinatra right before I perform, I'll sound just like Frank," he said. "If I listen to Nat Cole, I'll sing just like Nat. I have to be careful. Kevin Blanq, who produced my second album, was playing Louis Prima for me just before the sessions, and I had to tell him to stop."

I spoke with Connick for only about half an hour. In a sense, though, I spent the entire day with him, listening to his two Columbia albums—his eponymous all-instrumental 1987 debut (FC-40702) and the recent *20* (his age at the time of the recording sessions last summer: FC-44369)—in the morning and attending his first show that night. I was trying to make up my mind about him. Though he had no way of knowing it (and not that any musician should care what a critic thinks), Connick had gotten off to a bad start with me even before we met. The first strike against him was his pose on the first album's cover: dolled up in a big-shouldered polka-dot jacket, with glossed lips and slicked-back hair, he looked like he was supposed to be Pee-wee Herman's Dream Hunk. It was impossible to take him seriously as a jazz musician. (Upon seeing him in person, you realize that the mousse abuse was probably his idea, though the polka dots and lip gloss might not have been. If what I hear is true, jazz musicians on major labels these days control every aspect of their albums, except for the cover art.)

Strike two against Connick was the very fact that he had landed a Columbia contract even before arriving in New York, a city teeming with seasoned pianists who are lucky if they get to record for badly distributed independents. Pianistically, Connick didn't know who he was. On his own compositions, he was bland, fleet, and impressionistic in the manner of acoustic Herbie Hancock and Chick Corea, but without their finesse. On standards, he eschewed these contemporary influences for older, more idiosyncratic ones, and this impressed some conservative reviewers. But it seemed to me that Connick was being praised for his good taste in pianists (the New Orleans bluesmen James Booker and Professor Longhair, as well as a variety of jazz greats, including James P. Johnson, Willie "The Lion" Smith, Earl Hines, Duke Ellington, Art Tatum, Erroll Garner, and Thelonious Monk), rather than for his success at harnessing them into anything resembling a personal style (a typical solo was a jumble of Hines's tremolos, Monk's silences and dissonances, Longhair's rhumba cross-rhythms, Garner's stoptime chords and pianoforte runs, and Connick's own cute but carelessly executed boogie woogie and stride).

In a jazz marketplace in which youth is a commodity as well as a demographic (think of Wynton Marsalis), Connick has the considerable advantage of being young. He's from New Orleans, the son of a four-term incumbent district attorney, and a buddy of Wynton and Branford Marsalis (he has, in fact, studied with their father Ellis, and been produced by their younger brother Delfeayo). He's good feature material, and I thought that was all he was—until hearing him sing Harold Arlen in Philadelphia last summer.

As a singer, Connick captured Arlen's verve. In a nice bit of business copied from Armstrong, he added extra *ssss*'s to words, exploiting the effect for rhythmic emphasis. But in lieu of Armstrong's gravel, he delivered both Arlen's swing tunes and his ballads in the kind of clear, high, supple baritone that conveys a lovehurt more associated with tenors—a *Sinatra* baritone, in other words, as many (including Connick himself) have pointed out. To me, Connick sounds more like Bobby Darin (another Frankophile), though the resemblance is coincidental (he says he never listened to Darin until I first made the comparison in my *Philadelphia Inquirer* review of his Arlen concert). It's a question of vibrancy and timbre, but mostly a question of phrasing. Connick may not sing rock 'n' roll, and for that matter, neither did Darin (at least not willingly or very convincingly). But like Darin's, his relaxed delivery acknowledges the way we think, speak, and, most of all, sing differently as a direct result of rock. Significantly, Connick learns most of his songs from sheet music, rather than from records by older singers. He brings his own generation's accents to them.

The six vocals (out of eleven tracks) on *20* confirm the good impression Connick made singing Arlen in Philadelphia. His duets with an off-key Dr. John ("Do You Know What It Means to Miss New Orleans") and a bored Carmen McRae ("Please Don't Talk About Me When I'm Gone") are neither here nor there (the dreaded CBS duet syndrome at work again), but he sings "Imagination" beautifully, has fun with "Basin Street Blues" and "Do Nothing Till You Hear from Me," and turns in a strikingly original interpretation of Arlen's "If I Only Had a Brain"—the Scarecrow's song from *The Wizard of Oz*, recast as an overachieving yuppie's heartbroken ballad.

The obvious question is why did he wait until his second record to sing, given that that's where his real talent lies. "I wanted to establish myself as an instrumentalist first," he told me. "Not to put myself on Nat Cole's level, but he became so popular as a singer that most people were unaware that he was also one of the great jazz pianists. Same thing with Louis Armstrong. You ask people from my generation about him, and they say, 'Oh, yeah, that guy who sang.' I didn't want that to happen to me. It doesn't matter how good or bad a singer or piano player I am. If you sing, people think of you as a singer. Another reason I didn't sing on the first album was that I felt I wasn't ready yet. I was singing in a very gruff, New Orleans way, and doing a lot of damage to my throat." He illustrated by croaking a few bars of "If I Only Had a Heart" in a voice like Tom Waits's. "Then I tried to croon." He illustrated by singing a few bars of "Imagination" in a smarmy lounge style. "Finally, I hired Marion Cowings, a great legitimate singer, as a vocal coach, and he convinced me that I was better off singing in a voice like my natural speaking voice. I'll sing a lot more at the Algonquin than I would on a jazz gig. I'm not a cabaret musician, but a lot of the songs I do appeal to the sort of people who love cabaret. In a jazz club, I'd stretch out more, and play a lot more Monk, Duke, Miles, and Wayne Shorter—and some of my own tunes from the first record, which I don't think would go over well in a cabaret."

You never know. After Connick's first set, a woman at my table said she wished he'd played "more piano." In response, I blurted out, "I wish he'd sung more," which struck even me as strange, given that Connick sang on seven out of nine numbers. What I think I meant was that I wished he'd played *less* piano between vocal choruses. A fringe benefit of Connick's singing is that self-accompaniment imposes some restraint on his spendthrift piano style. From my point of view, it was the spare, diminished chords he played behind himself on "Imagination," not his frantic boogie-woogie pumping, that amounted to his best piano work of the night.

Because I enjoy Connick's singing so much, I'd rather hear him in a cabaret than in a jazz club. Even so, I'm not sure the Algonquin is good for him. Connick has an irresistible stage manner. A lanky six-footer, he comes across like Keith Carradine in an Alan Rudolph movie about jazz: a fascinating combination of talent plus guile (fascinating because the guile has the unexpected result of mitigating his overconfidence—it rises, I think, out of Connick's secret fear that he's not really as wonderful as he's certain everybody else thinks he is). But especially when leading a sing-along of "Shake, Rattle, and Roll," or joking about his fly being unzipped, Connick's youth and southernness are in danger of serving the same function at the Algonquin that Bobby Short's blackness serves at the Carlyle—that of reassuring smart-set patricians that they are *so* receptive to social change. Just as I wish that Connick would sing more in jazz clubs, I also wish he'd play more jazz (including the new songs he's writing with the New Orleans bassist Ramsey McLean— "contemporary songs, but not about AIDS or George Bush or anything like that") at the Algonquin's cabaret, even at risk of losing part of the audience, and even at risk of losing me. A friend of mine who's writing a book on the subject defines a child prodigy as "a child proficient at one adult task." Connick is a former prodigy, and the adult task at which he was proficient was showmanship. The ability to wow a crowd is a valuable asset for a performer of any age, but this twenty-one-year-old is already a master at that. It's time for him to develop other skills.

(*January*, 1989)

ONWARD

"When [I] first opened at the hungry i in San Francisco [in 1953], I used to do about fifteen minutes. . . . Since then, a lot of people think I've lost my discipline, and they say . . . 'Why does it take you an hour to cover a subject?' Believe it or not, it's not me. It's that there wasn't that much wrong in the country then and you could cover it a lot quicker," Mort Sahl quipped in *Heartland*, his 1977 memoir.

Mort Sahl on Broadway! the political satirist's 1987 one-man show at New York's Neil Simon Theater, ran just under *two* hours. But don't jump to conclusions about Sahl's current opinion of the state of the union. Sahl doesn't play favorites. Reagan is "a nine-to-five president—make that nine-to-ten. . . . George Washington could not tell a lie, Richard Nixon could not tell the truth, Ronald Reagan cannot tell the difference." The surplus of Democratic presidential hopefuls "in the Kennedy mold" causes Sahl to wonder, "Is there anyone in the party who's not like Kennedy? Yes—Teddy Kennedy." He confides that he's not against funding the contras as long as it's for "humanitarian" purposes, "like a suite for Arturo Cruz at the Ambassador Hotel when he does *Nightline*." Turning his attention to Hollywood, he tells how, with tears in his eyes, he threw away all the popcorn, salad dressing, and spaghetti sauce in his refrigerator after an ideological disagreement with Paul Newman. But it isn't only Hollywood's liberal cause celebs who amuse him: Charlton Heston is "a cerebral fascist," who thinks, "'I, unfortunately, won't live to see it, but someday my grandchildren will live under a military dictatorship.'" If Heston were "more perceptive, he'd be happy now."

This is funny stuff, but, except for the rip at Heston, no more seditious than the gentle tweaking that Johnny Carson regularly gives the high and mighty. It's impossible to tell from these nonpartisan one-liners what Sahl's own politics are—which is all right for Carson and Bob Hope, who depend on writers and give you the feeling that comedy is a job, but not all right for Sahl, who's too egotistical to let anyone else write his material, and who used to entertain audiences with the news that the military and the CIA were running the country.

That was the message of *Heartland*, which was written during a period when few club owners or television producers would touch Sahl, for fear that he might read aloud from the Warren Report. This was the low point of Sahl's career, and what a drop it must have been for someone who had reached such pinnacles as co-hosting the 1960 Academy Awards (with Hope, Jerry Lewis, Laurence Olivier, and Tony Randall) and being profiled in *The New Yorker* and featured on the cover of *Time* that same year.

Did Sahl fall or was he pushed? In *Heartland*, he claimed he'd been blacklisted as punishment for going to work for New Orleans D.A. Jim Garrison on the Kennedy assassination and daring to speculate about the CIA's involvement. Sahl may have been right, just as Garrison may have been. But those who shunned Sahl would probably argue that he had become a paranoid and crank; significantly, he did find work on two-way talk radio, where being a paranoid and crank is among the job qualifications.

My own guess is that Sahl's career would have gone into reverse even without Garrison. Sahl was the most innovative comedian to gain access to a mass audience in the comedy-happy fifties (you'll have to take my word for it, because the albums that would prove it are long out of print). Lenny Bruce was still doing *shtick* when Sahl began walking onstage with a newspaper, extemporaneously riffing on the headlines. A cool-jazz buff, Sahl derived from musicians not only his timing, but also their habit of traveling light—schnooky dialects and comic personas like the ones Bruce was still hiding behind would have been excess baggage for Sahl, who spoke to audiences in his natural voice, that of a pseudo-intellectual positive he was smarter than the assholes running things. In 1960, Richard Nixon, long one of Sahl's favorite targets, and never more eager to prove that he had a sense of humor than in an election year, told *Time* that Sahl was "the Will Rogers of our time." Sahl's comeback: "Rogers . . . impersonated a yokel who was critical of the federal government. . . . I impersonate an intellectual who is critical of the yokels who are running the federal government. Other than that, we're similar in every respect."

Nixon's quote is reprinted in the *Playbill* for *Mort Sahl on Broadway!*—but without Sahl's rejoinder. As Nixon's blessing might suggest, Sahl got away with as much as he did because the people he skewered assumed he was only kidding, and maybe they were right. He thought of himself as "the loyal opposition," critical of whoever was in power, liberal or conservative. The bigwigs saw him as a court jester, and they dropped him when his humor turned sour (it was one thing to rib JFK about his father's wealth, another to make wholesale accusations about his murder). At the same time, Sahl's show-biz veneer and in-ness with Hugh Heffner and various New Frontiersmen made him an unlikely candidate to be speaking truth to power, in the eyes of the audience that was coming to prize irreverent humor most—disaffected, vaguely leftist youth. The biggest barrier for this audience was Sahl's misogyny, which was always there, but which became more virulent in response to feminism. Here was a man who, as late as 1977, was capable of writing: "When I went to New York in the '50s, you had to be Jewish to get a girl. In the '60s you had to be black to get a girl, and now you have to be a girl to get a girl." That's at least funny—the sort of joke feminists are likelier to laugh at than "sensitive" men, if only because no man has ever experimented with homosexuality for ideological reasons. But what about this: "A writer and I were talking about a picture called *How to Make Love in Three Languages*. We idealized the last scene, in which a girl who insists on being an actress falls in love with an analyst. She tears up her Social Security card, throws it in the Trevi Fountain, and decides to be a woman, which is the only decision for a woman to make." In the seventies, politics gradually became as much a question of values as issues, and Sahl stubbornly refused to make the transition. He was soon left with no audience at all.

Now Sahl is back and Broadway's got him, which is, respectively, the good news and the bad. The audience for the matinee performance I attended was considerably smaller but otherwise not much different than the middle-aged-to-elderly and moneyed-looking crowd lined up around the corner for *Cats*. A typical Broadway crowd, in other words, which Sahl seemed to interpret to mean a mostly Jewish crowd. Although Sahl is both Jewish and a comedian, he's never come on like a *tummler*, so it was shocking to watch him courting favor with easy put-downs of Vanessa Redgrave and inanities about second-generation Jews who don't speak Yiddish. Sahl's only previous show on Broadway was the 1958 revue *The Next President*, the title of which, although it referred specifically to who would be Eisenhower's successor in Washington, was a dead giveaway that Sahl was (as fellow egoist Norman Mailer once said about himself) running for president in the privacy of his mind. Now that Sahl is back on the campaign trail, as it were, Jackie Mason's successful comeback has persuaded him of the wisdom of going after the Jewish vote.

Without the cocksure ebullience he possessed in the fifties or the bilious anger that ate away at him later, Sahl is in danger of becoming Mark Russell with sharper incisors and a hipper delivery. But after the dues he's paid, who has the heart to accuse him of selling out? Besides, he can be outrageously funny, especially when he isn't joking—when he addresses sexual politics and betrays his rancor at the way cultural values have shifted against him. You know you shouldn't be laughing, but you can't help yourself, and part of it is the ironic realization that it took a self-described "puritan" who wouldn't dream of using profanity on stage to identify our new taboos. "The new woman doesn't want to get married," he says, and adds that that's fine with him, because "men have never wanted to get married. So we now live in a country with two-hundred-and-forty-million people and nobody wants to get married and have children— except for the occasional Catholic priest." Watching his wife dance with George Bush at the White House, Sahl is troubled: "and not only because it was the first time I saw her smiling since she met me," but also because he suspects that he has more in common with Bush than he would with a liberal Democrat vice-president "who would probably be saying to her, 'Why don't you leave that crumb? After all, *you're* a person, too. Go back to college and study the social sciences. You can help lesbians to adopt children!'"

Sahl's wife—China Lee, a 1964 *Playboy* centerfold—gets her turn around the floor with Bush during a discursive but brilliantly sustained narrative that reveals Sahl to be a monologist as well as a comic—sort of an outer-directed Spalding Grey. Sahl is in his office at the Warner Brothers lot, where he makes a living writing screenplays that get optioned but never filmed (and where, to stay in good graces, he must "maintain the illusion of still being a liberal"). He's going through his mail, and there at the bottom of the pile is an invitation to a state dinner in honor of Israeli Prime Minister Yitzhak Shamir. Paul Newman, Warren Beatty, Dustin Hoffman, and other liberal friends urge him not to accept, but his wife thinks he should. Mort is undecided. "'You don't know how to be happy . . .'" China nags him: "'You're not happy when they accept you, you're not happy when they don't accept you. You're not happy when you're inside, you're not happy when you're outside. . . .'"

Sahl finally decides to go: "It's only a dinner." In the reception line, he's impressed that the president has a different joke for each couple. After exchanging pleasantries with Ron and Nancy, Mort meets Shamir, who, upon learning that Sahl is an entertainer, instructs him to beseech his audiences to urge their government not to sell arms to the Saudis ("Israelis are lots of fun at parties"). Later, he overhears an argument between Shamir and Caspar Weinberger over whether the U.S. is indeed already instructing the Saudis in the use of U.S.-supplied weaponry. He informs Reagan, who dissolves the tension with an anti-Arab joke. As the President and First Lady stand waiting with Mort and China for their limo, Ron tells Mort how much he's looking forward to his retirement, and charms him with an apocryphal story about Jimmy Carter and Menachem Begin meeting at the Wailing Wall on the anniversary of the Camp David accords.

This is Sahl's *Swimming to Washington*. From the retelling, it might seem as though he's fallen in love with Reagan for the reason he once fell head over heels for JFK—he's a sucker for a leader with a sense of humor. But the story isn't really about Reagan. It's about Sahl's ambivalent relationship to power. He knows that, as a political satirist, it's his calling to reveal that the emperor has no clothes. But when actually in the emperor's presence, he's like a painter who becomes infatuated with his nude model. Not having an audience must have been an ordeal for Sahl—he gives the impression of being the talkative sort who, if all else failed, would consider becoming an alcoholic just for the pleasure of sounding off to strangers in bars. But the worst part of being considered washed up must have been that it cut him off from his best source of material: the behind-the-scenes machinations of the Hollywood and Washington power elites. His inclusion of China Lee's harangue seems intended to show that, given seventeen years of marriage, even an Asian-American Playmate can become a yenta. But it also shows that Lee has her husband's number. Presidents—can't live with 'em, can't live without 'em. He's not happy when they accept him, but he's not happy when they don't accept him, either, because he's learned that he needs them more than they need him.

Like most one-man shows, *Mort Sahl on Broadway!* is negligible as theater. But Sahl himself is the stuff of drama. Despite raves in the *Times* and *The New Yorker*, the show closed on November 1, exactly as first announced: it wasn't held over, as Sahl confidently told the *Times* it would be. But it wasn't a flop, either, and the middling result hardly seems fair—someone with Sahl's sense of self-importance deserves either complete redemption or crushing failure. I don't know where he goes from here, but I'm hoping it's back up.

(November, 1987)

combos, movements, issues, and isolated events

POSTMODERNISM, 1924

George Gershwin's *Rhapsody in Blue*—which was commissioned by the danceband leader Paul Whiteman and given its premiere by Whiteman's Orchestra (with the composer on piano) at New York's Aeolian Hall on February 12, 1924—occupies a tenuous niche in the modern orchestral repertoire. Essentially a piano concerto, it's a bounding showpiece for soloist and orchestra, a real crowd-pleaser. Counting "pops" concerts, it's probably performed more often than any other work by a twentieth-century American composer. But it's usually treated as a diversion, a relief from weightier concert fare, because the Europhile classical music establishment still regards any "serious" work indebted to American popular music as a bastard offspring.

Meanwhile, the combination of *Rhapsody in Blue*'s white-tie-and-tails origins and its misrepresentation as fodder for the likes of Arthur Fiedler has alienated the jazz and pop audiences for whom Gershwin undoubtedly intended it as an early valentine. Whiteman's Aeolian Hall concert, the first such intrusion into a classical music sanctuary by an American dance band, discounting an overlooked 1912 Carnegie Hall program by James Reese Europe and his Clef Club Symphony Orchestra, was designed to make "an honest woman out of jazz"—a sentiment that has done nothing to endear Whiteman or Gershwin to those who hear nothing particularly whorish about jazz unwed to the symphony. To make matters worse, from 1951 to 1957, in an early demonstration of television's propensity for reducing everything to glucose, a candied reading of *Rhapsody in Blue*'s *andantino moderato* penulti-mate movement (its slowest passage, and the only one dominated by strings) served as the theme for NBC's *Goodyear Playhouse*. For baby boomers weaned on TV, Gershwin's *Rhapsody* thus came to epitomize the genteel tradition that Elvis Presley trampled on the *Ed Sullivan Show*.

In 1924, *Rhapsody in Blue* climaxed an evening that also featured works by Zez Confrey, Ferde Grofé, Victor Herbert, and Rudolf Friml. Three years ago, in commemoration of the sixtieth anniversary of the Aeolian Hall concert, Maurice Peress mounted a reconstruction of Whiteman's "Experiment in Modern Music" at New York's Town Hall. Now that there's a two-record digital studio version of Peress's re-creation—*The Birth of Rhapsody in Blue* (Musicmasters MMD 20113X/14T)—it's clear that the *Rhapsody in Blue* we've grown to love or detest is a pale echo of what was played that night in 1924, for an enthusiastic audience that included such classical music dignitaries as Jascha Heifetz, Fritz Kreisler, Sergei Rachmaninoff, and Leopold Stokowski.

In his program notes for the premiere (reprinted in their entirety on the Musicmasters release) Gilbert Seldes, the author of *The Seven Lively Arts*, a ground-breaking polemic on behalf of America's indigenous popular arts, predicted that *Rhapsody in Blue*, even in its augmented dance-band instrumentation, would appeal to symphonic conductors. "[It will] probably need rescoring," Seldes cautioned, "but the saxophone, which has been used ever since Meyerbeer in serious music, need not be exiled." Of course, saxophones *have* been banished from most subsequent orchestrations (including the symphonic score that Ferde Grofé prepared a few months later), along with the banjo, an instrument considered totally out of bounds in longhair music.

But Peress, a Seldes disciple who is becoming our most valuable archeologist of vintage pop (he also conceived the American Music Theater Festival's 1984 revival of Gershwin's *Strike up the Band*, and was an important contributor to the same company's 1986 production of Duke Ellington's previously unperformed *Queenie Pie*), has restored the déclassé instruments to the position of dominance they enjoyed in Grofé's original dance-band orchestration, and the difference in rhythmic vitality and harmonic shimmer is astonishing. It's now possible to ascertain that one of Gershwin's appointed duties was to show off the splendors of Whiteman's reed and woodwind section, which featured Hale Byers, Donald Clark, and Ross Gorman, all of whom were virtuosos on at least three different horns. And that dreaded slow movement is far more palatable with Eddy Davis's rambunctious but apt banjo strumming behind the violins like a rhythm guitar.

To say that this *Rhapsody in Blue* is different from any other on record isn't critical hyperbole, merely a statement of fact. In addition to restoring the banished instruments (and thereby reclaiming the work for jazz), Peress has also reinserted piano and orchestral passages excised, for reasons of space, from the presumably definitive recording that the composer made with Whiteman shortly after the premiere. But Peress's fidelity to *Ur*-text only partly accounts for his interpretation's appeal. The piano soloist Ivan Davis turns in a reading that is exemplary by any standard, but especially crafty in dealing with Gershwin's tricky retards. The opening clarinet glissando to high B-flat, notorious for tempting even virtuosos to shrillness, is given the jazz grit and vibrato that Gershwin probably intended (in the fashion of a jazz composer, he wrote the figure specifically for Gorman after hearing him play something very much like it at a Whiteman rehearsal), and the tempos are blessed with the jazz swank that one imagines the Whiteman Orchestra was striving for. Perhaps they are blessed with too much swank: if Ellington's and Fletcher Henderson's orchestras weren't yet swinging in 1924—and records from the period prove that they weren't—it's doubtful that Whiteman's was. But this minor gaffe, apparent not only on *Rhapsody in Blue* but throughout the double album, makes the set far more agreeable to contemporary ears than it might otherwise be. And in a miraculous transformation, Gershwin's relationship to jazz no longer seems derivative, but prescient in regard to James P. Johnson's "Yamacraw," Ellington's "Creole Rhapsody," and Barney Bigard's wrap-around clarinet gliss on "The New Black and Tan Fantasy."

If Peress's transcriptions are an accurate indication, *Rhapsody in Blue* outclassed everything else performed at Aeolian Hall that night, but the other stuff was pretty delightful, however minor. What we have here is a veritable Caucasian kaleidoscope, with nearly every important white pop composer of the early twenties endeavoring, not always successfully, to come to terms with black dance rhythms. Surprisingly, Edward MacDowell's tearjerker "To an Irish Rose" and Rudolf Friml's milky "Chansonnette" (unfortunately retitled "Donkey Serenade" after being saddled with lyrics), two of the pieces you'd expect to have dated most, are among the set's freshest sounding works. It would be easy to dismiss Zez Confrey's player-piano novelties (adroitly reproduced here by Dick Hyman), Victor Herbert's sentimental serenades, and Grofé's coy variations on classical marches and arias as period kitsch—but only after acknowledging what an unprecedented period it was, and what noble kitsch it produced. This was a time of flux in American music, as composers like Gershwin, Herbert, Confrey, and Grofé heeded both New World impulses and Old World propriety. Although most of this music registers as quaint today, one could make a case for it as postmodern in its willingness to toss everything into the Cuisinart. Today's rock and classical mix-and-match types should show such savvy in knowing where to shop for fresh ingredients.

Whiteman's rabble following and patrician aspirations made his orchestra a crucible for the healthy give-and-take that went on between high art and popular culture in the 1920s, before pop, jazz, and classical hardened into discreet categories. Just as Hyman's substitution of a trio of lovely impressionistic études for a lost comedy selection should do wonders for Confrey's tainted reputation, so should the chimerical textures that Peress draws out of the original band scores go a long way toward redeeming Whiteman's good name. After acknowledging Whiteman's taste in sidemen, standard jazz histories (Gunther Schuller's *Early Jazz* excepted) vilify him as a white usurper, perpetuating the backlash from his having erroneously been coronated "The King of Jazz" in the 1920s, when few perceived that the brash new music would prove to be an enduring Afro-American art form rather than a passing dance craze. To judge from the pronouncements of his manager, Hugh C. Ernst, in the Aeolian Hall program notes, Whiteman's understanding of both jazz and pop was rudimentary, his approach didactic and condescending. He intended, for example, to show how a medley of Irving Berlin tunes might assume a more "dignified" posture when treated symphonically; by "jazzing" the final chorus of "Whispering," he meant to demonstrate "how any beautiful selection may be ruined" by such wanton indiscretion. Hindsight proves what I suspect Whiteman's audiences already knew, that Berlin's melodies had an innate dignity all their own, and that "jazzing" a standard involved more than playing hot and loose (though, in all fairness, Whiteman shouldn't be chastised for failing to anticipate "Groovin' High," the bop anthem that Dizzy Gillespie based on "Whispering's" higher intervals some twenty years later). But for all his shortcomings, Whiteman approached the task of eroding the boundaries between the highbrow and the lowbrow (the ongoing mandate of American music in the 1980s, with most of the vigorous initiatives still coming from the lowbrows) with a zeal demonstrated by no other white bandleader before or since. His Aeolian Hall concert was a landmark event in American music for that reason. So, in its own way, is Peress's loving re-creation—and not merely for its lustrous, slightly antic *Rhapsody in Blue*.

(*April, 1987*)

SMITHSONIAN WORLD

SOMETHING TO REMEMBER THEM BY

The decline of the operetta as the model for the Broadway musical, the displacement of belters by crooners when improvements in amplification rendered irrelevant the question of a singer's projection, the impact of jazz and blues on white pop, the gradual and perhaps inevitable gentrification of a once virile idiom as the popular songs from the twenties, thirties, and forties—no longer "popular" in any true sense—fell under the guardianship of performers inclined to treat them as art songs . . . these are a few of the themes of the Smithsonian's *American Popular Song: Six Decades of Songwriters and Singers* (Smithsonian Collection of Recordings R 031 P 7-17983), a lavishly packaged set of seven records (and a 152-page booklet) surveying the achievements of American popular composers, lyricists, and vocalists in the fifty years between Victrolas grinding out ragtime and transistor radios blaring out rock 'n' roll.

Many of the songs and performers chosen by the producer J. R. Taylor and the consultants Dwight Blocker Bowers and James R. Morris flout conventional wisdom. The singer most prominently featured, for example, is Fred Astaire, with nine selections (Frank Sinatra, Ella Fitzgerald, Tony Bennett, Judy Garland, and Nat "King" Cole have five apiece). The composer taking top honors is Harold Arlen, represented by an even dozen of his songs ("Over the Rainbow" and "One for My Baby" surprisingly not among them)—two more than Richard Rodgers, three more than George Gershwin, four more than Irving Berlin, and six more than Cole Porter, Jerome Kern, or Hoagy Carmichael. This elevation of Astaire and Arlen, though unexpected, is astute. Astaire's singing, no less than his dancing, for which he is most celebrated, epitomizes equilibrium and nonchalance—states of being all pop singers strive for, or should. Arlen, who has never enjoyed as much cachet as Gershwin, Kern, Porter, or Rodgers—perhaps because he lacks their pedigree as men of the theater—is generally extolled as a craftsman rather than as an innovator. Yet he's given us more great songs than any other American songwriter, and he was more successful than any of his contemporaries in catching the sass and fervor of blues and jazz. Because Astaire and Arlen spread their favors around, they cover a lot of *American Popular Song*'s territory all by themselves. Astaire glides through songs by Berlin ("Cheek to Cheek," "Isn't This a Lovely Day," and the unbowdlerized "Puttin' on the Ritz," which also features a tap chorus), Gershwin ("They Can't Take That Away from Me" and "Fascinatin' Rhythm," the latter a duet with sister Adele, with the composer at the piano), Kern ("A Fine Romance"), Porter ("Night and Day"), Arthur Schwartz ("By Myself"), and Johnny Mercer ("Something's Gotta Give"). Arlen's collaborations with lyricists as different in sensibility as Mercer, Ted Koehler, E. Y. "Yip" Harburg, Truman Capote, and Ira Gershwin provide choice material for several generations of singers, including Jack Teagarden ("I Gotta Right to Sing the Blues"), Lena Horne ("Stormy Weather" and "As Long as I Live"), Judy Garland ("Get Happy" and "The Man That Got Away"), Ella Fitzgerald

("Blues in the Night"), Mabel Mercer ("My Shining Hour"), Mel Tormé ("When the Sun Comes Out"), Tony Bennett ("A Sleepin' Bee" and "Last Night When We Were Young"), Joe Williams ("Come Rain or Come Shine," with the Count Basie Orchestra), and Aretha Franklin (a delightful gospel rave-up of "Ac-cent-tchu-ate the Positive"). *

Ubiquitous though they may be, Astaire and Arlen hardly monopolize *American Popular Song.* There are other riches here, beginning with Sophie Tucker's 1910 Edison cylinder recording of "Some of These Days," a likably bravura performance, but rhythmically stilted in contrast to the track that follows it: Bessie Smith's supple declamation of Berlin's "Alexander's Ragtime Band," with backup from Fletcher Henderson, Coleman Hawkins, Joe Smith, and Jimmy Harrison. Smith's phrasing was a portent of things to come. Although the majority of the songs included in this set were written in the thirties, and none after 1955, no fewer than forty-six of the 110 selections were recorded after that arbitrary cutoff date. This testifies to the longevity of the best popular songs, and to the role of the long-playing album in assuring that longevity—artists need songs to fill up space. Among the most diverting performances are those by vintage recording stars whose names are hardly mentioned anymore: the prototypical crooner Gene Austin (an unsaccharine "My Melancholy Baby"), the versatile twenties coloratura Marion Harris ("After You've Gone," "I Ain't Got Nobody," and a flawless "The Man I Love"), and the band singer Irene Taylor (a snazzy "Willow, Weep for Me," with the Paul Whiteman Orchestra). The four Bing Crosby selections are well chosen. They reveal his debt to Al Jolson and vaudeville, but show what distinguished him from his predecessors: the forthright sensuality of his phrasing, and his feeling for jazz rhythm (the violinist Joe Venuti and the guitarist Eddie Lang are assumed by the producers to be among Crosby's unidentified sidemen on the stirringly hoarse 1931 broadcast version of "I'm Through with Love"). Sinatra's "I've Got You Under My Skin," Garland's "The Man That Got Away," and Holiday's "You Go to My Head" remain powerful no matter how many times you've heard them, perhaps all the more so for the confusion of firsthand and inherited memories they call into play. But the anthology's major revelation is "Dancing on the Ceiling," by the underappreciated Jeri Southern—a gossamer and oddly sibilant 1952 rendering of Lorenz Hart's lyric that achieves confidentiality without coyness or guile. It's an

---

*Arlen and Astaire worked together once, but there's no evidence of their collaboration on *American Popular Song. The Sky's the Limit* (1943) isn't remembered as one of Astaire's better movies, though its score included Arlen and Mercer's "My Shining Hour" and "One for My Baby."

inspired selection, especially in light of Sinatra's justifiably famous recording of the tune, commonly regarded as the definitive interpretation.

American Popular Song also has a share of baffling choices: overzealous attempts to tell a story in song by the arrangers Lalo Schifrin and Pete Rugulo (Sarah Vaughan and Nat Cole are their respective victims); feckless ballads by the male ingenues Russ Columbo, Buddy Clark, Johnny Mathis, Gordon MacRae, and the young Perry Como; and arch recitations by Portia Nelson, Elaine Stritch, Eileen Farrell, and Barbara Cook, cabaret divas who bring the curse of gentility to everything they do. The many sins of omission are even more inexplicable. Two numbers each by Billie Holiday and Ethel Waters hardly seem sufficient, but they at least fare better than Louis Armstrong, Mildred Bailey, Connie Boswell, Billy Eckstine, and Jack Teagarden, each of whom is limited to one appearance; and Lee Wiley, Ruth Etting, Maxine Sullivan, Johnny Hartman, Helen Humes, Jimmy Rushing, and Ray Charles, who are ignored altogether. The compilers have a regrettable aversion to "list" songs of the type popular in the 1930s. "How About You," "These Foolish Things," and "Thanks for the Memory" are admittedly irresistible for their wordplay rather than their melodies, but irresistible nonetheless for the way they translate romantic longing into imagery of everyday objects: potato chips, moonlight, and motor trips; a cigarette that bears a lipstick's traces; stockings in the basin when a fellow needs a shave. The vital contributions of journeyman songwriting teams like Bert Kalmar and Harry Ruby, Ralph Rainger and Leo Robin, and Jules Styne and Sammy Cahn are given short shrift or allowed to go unmentioned. Duke Ellington, a great composer but an indifferent songwriter is well represented, but Fats Waller, Eubie Blake, and James P. Johnson are missing. And where is Matt Dennis? After all, the question of whether his "Angel Eyes" is a great song is academic; it is when Sinatra sings it, and his touchstone recording of it should have been included, along with Louis Armstrong's shocking interpretation of Waller's "Black and Blue."

American Popular Song is the stepchild of the late Alec Wilder's revisionary 1972 book of the same name. Although Wilder, a gifted songwriter but not a very prolific one, is represented only twice, his sensibility pervades the selections. Like this frustrated "serious" composer, Taylor, Bowers, and Morris prize melodists over lyricists, and relegate vocalists to a custodial position—the composer is their *auteur*. But like commercial filmmaking, pop songwriting during the half century covered by this collection was often collaborative. Lorenz Hart's verse, for example, gave Richard Rodgers' early melodies their cosmopolitan sparkle, and not the other way around—if we're to draw conclusions from the humdrum melodies Rodgers later appended to the chamber-of-commerce pronouncements of Oscar the-corn-is-as-high-as-an-elephant's-eye Hammerstein II. Likewise, even though Jerome Kern is said to have treated his lyricists as paid underlings, Dorothy Fields and Johnny Mercer brought to songs like "A Fine Romance" and "I'm Old Fashioned" a colloquial verve missing from most of Kern's portfolio, whatever its other charms. Even Irving Berlin and Cole Porter, who wrote their own lyrics, frequently tailored their melodies to flatter the singers (and, in some cases, the vocally restricted singing actors) who would be introducing them in films and shows. With the possible exception of Kern, the great American songwriters were seldom as intractable as their latter-day savants.

Wilder, despite his enthusiasm for Billie Holiday, Mildred Bailey, and Lee Wiley, was ambivalent about vocal improvisation, saving his highest accolades for singers who took few liberties. Still, I find it surprising that jazz singers are so poorly represented in the Smithsonian box, given that Taylor, a former jazz critic and a good one, had a hand in the selection. It's largely been through the efforts of jazz singers, cavalier about melodic text though they may sometimes be, that the hit tunes of former decades have endured as standards. Although jazz instrumentalists turn up as accompanists on *American Popular Song*, it might have been fitting also to include a few strictly instrumental performances, such as Coleman Hawkins's "Body and Soul," Charlie Parker's "Embraceable You," Lester Young's "These Foolish Things," and Thelonious Monk's "Smoke Gets in Your Eyes," to show how pop songs have provided malleable raw material for harmonic investigation.

I'd also quarrel with the fatalistic 1955 cutoff date (when the barbarians arrived at the gates, from one point of view). Granted, the best tunes of the rock era have been great records rather than great songs. And granted, for the past thirty years, Broadway showtunes have had to delineate character and further plot, which explains why so few of them have had any currency off the stage. Yet "Send in the Clowns" from Stephen Sondheim's *A Little Night Music* and the title song from *Anyone Can Whistle* have managed to enter the standard repertoire, and so (with an assist from Sinatra) has "New York, New York," John Kander and Fred Ebb's anthem for the great Martin Scorsese film of the same name. These songs deserve places on the Smithsonian's honor roll, as do less frequently performed Sondheim oratorios as "Good Thing Going" from the ill-fated *Merrily We Roll Along*, "There Won't Be Trumpets," which was dropped from *Anyone Can Whistle* during the out-of-town tryouts, and "Losing My Mind," which Dorothy Collins sang so beautifully in *Follies*. Room might also have been found for such relatively recent examples of inspired hackwork as Jimmy Van Heusen and Sammy Cahn's "Call Me

Irresponsible" and Johnny Mandel and Paul Francis Webster's "The Shadow of Your Smile," and for such rock-era oddities as the Flamingos' haunting renovation of "I Only Have Eyes for You," the Spaniels' "Stormy Weather," Frankie Lymon's heartfelt renderings of Rube Bloom's "Fools Rush In" and "Out in the Cold Again," and the British postmodern crooner Bryan Ferry's fiendishly syncopated "These Foolish Things."

But enough. My differences of opinion with *American Popular Song* vanish whenever it's on the turntable. This is a one-of-a-kind collection of the songs and performances that ritualized mid-century American attitudes toward sex, courtship, infidelity, affluence, leisure, and aspiration. In the mid–nineteenth century, when Walt Whitman declared that he heard America singing, he was speaking metaphorically. But the sentiment is frequently taken at face value: America is assumed to find its clearest voice in robust carols of fellowship and labor (witness the periodic vogue for folk and ethnic musics among leftists, and the current ideological tug-of-war over Bruce Springsteen's "Born in the U.S.A."). The truth of the matter, though, is that Americans—at least since the invention of machines to do their singing for them—have generally eschewed celebrations of labor and fellowship in favor of escapist fantasies of the good life and plaints of loneliness and despair.

This isn't necessarily the evidence of decadence that hard-line critics of popular culture say it is. It was from the urbane and sometimes frivolous songs of Gershwin, Berlin, Rodgers, Kern, Porter, and Arlen (and from the equally frivolous stage and film musicals for which they wrote many of those songs) that those of us born on the bottom of the pile, or near the middle, first gained a clue to how the other half lived. These songs were the stuff of dreams, and, often enough, the music-makers were themselves living proof that dreams could come true—up-from-the-masses composers and lyricists and singers all helping to free American society from the restraints of the past. Tin Pan Alley was our assimilationist bohemia, a world-that-never-was where a Jewish immigrant like Irving Berlin could write best-loved songs for Christian and patriotic holidays, where white singers like Sophie Tucker and Marion Harris could sway to the rhythms of black composers like Shelton Brooks ("Some of These Days") and Turner Layton ("After You've Gone"), where homosexuals like Lorenz Hart and Cole Porter could write the lyrics that reminded fighting men of their girls back home, where slummers and social climbers could rub elbows, if only for the duration of a song. *American Popular Song* belongs in every record library, as much for the unofficial social history it conveys as for the musical pleasure it affords.

(July, 1985)

STRUGGLIN' WITH SOME BARBEQUE

In a few paragraphs, I'm going to argue with Martin Williams, which will be like raising my voice to Father. Unless I flatter myself about my lineage, he's going to recognize certain of my values as his own. I'm not Williams's only progeny. Although never as influential with a mass readership as Leonard Feather, Nat Hentoff, Ralph Gleason, or Whitney Balliett, Williams has long been the jazz critic from whom others in the field take their cues. Himself the heir of André Hodeir, Williams borrowed and expanded on the French critic's heretical thesis that rhythm—not harmony—was the force behind every significant innovation in jazz, including bebop. This is now something that every jazz critic accepts as gospel, as is the idea that jazz has an unbroken tradition from ragtime on, with a period of consolidation following each stylistic upheaval—an argument that Williams first advanced in short *down beat* record reviews in the late 1950s, while never straying too far from the album at hand.

Broad but discriminating in his enthusiasms, Williams introduced a degree of empiricism to a field in which passionate likes and dislikes had long been deemed sufficient journalistic credentials. Although sociology (the purview of Hentoff, his fellow editor at the short-lived *Jazz Review*) was never his strong suit, Williams never downplayed the crucial issue of race. And in his writings about film, television, and comic strips as well as jazz, he confronted the dialectic that mocks all critics of the "popular" (as opposed to the "fine") arts: the frequent lack of correspondence between popularity and merit. Hardly one of jazz journalism's finer imagistic prose stylists, he nonetheless achieved a clarity of expression others would do well to emulate. Balliett is the father of show-offs who want their words to leap off the page, though they'd probably ridicule his excessively refined tastes in both music and metaphor. Williams is the model for those of us who'd prefer our words to stay put, at least until we're positive they're the right ones for the job.

I've been discussing Williams's accomplishments in the past tense because he's ceased to be a full-time critic since going to work for the Smithsonian in the early seventies. But if anything, he's acquired greater eminence during his tenure in Washington. In fact, *The Smithsonian Collection of Classic Jazz*, released in 1973, might be Williams's greatest contribution to jazz discourse. This boxed, six-record survey of jazz evolution from ragtime to Ornette Coleman amplified *The Jazz Tradition*, Williams' definitive text, first published in 1970 (revised in '83). Criticism you could tap your foot to, *Classic Jazz* provided a framework for the reissue boom then already in progress, and a favorable climate for the jazz repertory movement just around the corner. But it also scored a semantic victory for modernism by rescuing the phrase *classic jazz* from New Orleans traditionalists and applying it across the board, to Charlie Parker and Ornette Coleman as well as to King Oliver and Jelly Roll Morton.

Understandably, *Classic Jazz* concentrated on the major figures, and if Williams' reckoning of who they were differed from the general consensus, that was part of the fun. Williams shares certain traits with Pauline Kael, the only film critic comparable to him in stature. Like her, he has massive blind spots. Just as she has little use for John Ford or Alfred Hitchcock, Williams has never warmed to John Coltrane, whom he severely underrepresented with just two appearances on *Classic Jazz* ("Alabama," plus "So What," with the Miles Davis Sextet) at a time when the late tenor saxophonist had jazz in a posthumous hammerlock. Still, this was one of those instances in which wrong was right: Williams correctly perceived that Coltrane impersonators were leading jazz toward a cul de sac, with their know-nothing mysticism, droning modality, and opportunistic black cultural nationalism. Besides, Williams's distrust of Coltrane illustrated the immunity to received opinion that blessed *Classic Jazz* (which didn't include "Giant Steps," "'Round Midnight," or "Sing, Sing, Sing") with a consistency of purpose no anthology chosen by consensus could hope to match.

Earlier this year, when the Smithsonian announced plans for a revised edition that would include seven albums instead of six, two possible rationales suggested themselves: one commercial (the new market opened up by digital remastering and compact discs); the other aesthetic (the two extra sides could be used to accommodate music recorded after the first release). Wrong on both counts. Like its predecessor, this edition of *The Smithsonian Collection of Classic Jazz* (Smithsonian Collection of Recordings R 033 P7-19477) is available on LP and cassette—not CD. The tracks don't seem to have been digitally remastered, which may be just as well, given that digital equipment has a way of reading drum sizzle as surface noise, and that the original edition sounded fine to begin with. But the real surprise is that only two of the added performances were recorded after 1973: Sarah Vaughan's "My Funny Valentine" and the World Saxophone Quartet's "Steppin'."

My vague discontent with the new *Classic Jazz* centers around what's missing from it, not what's there. Louis Armstrong, Duke Ellington (eight selections each), Charlie Parker (six, plus two alternate takes), Jelly Roll Morton, Thelonious Monk (five each), Ornette Coleman, and Count Basie with Lester Young (three each) dominate the collection, as well they should.

Williams has jettisoned thirteen tracks from the original, presumably after realizing they were superfluous (for example: Billie Holiday's "All of Me" from 1941, similar in spirit to "He's Funny That Way" from four years earlier, has given way to a version of "These Foolish Things" from 1952, when Holiday had gained dramatic bite in compensation for her loss of lilt). This strategy backfires in a few instances. Robert Johnson's "Hellhound on My Trail" belongs here alongside Bessie Smith as a reminder that the blues retained their original identity even as they gave shape to jazz. "Creole Rhapsody" deserved its former place of honor as the first of Ellington's extended works, and Monk's stripped-down quintet arrangement of "Smoke Gets in Your Eyes," which didn't make the final cut, might have dramatized his powers of distillation even more effectively than his solo version of "I Should Care."

In all, Williams has added a staggering twenty-one tracks not on the original. Most of these are beneficial substitutions: Lennie Tristano and Lee Konitz's "Subconscious Lee" for "Crosscurrents," Bud Powell's "Night in Tunisia" for "Somebody Loves Me," Charles Mingus's "Haitian Fight Song" for "Hora Decubitus," and so on. Other additions feature performers overlooked on the first edition: Jimmy Noone, Django Reinhardt, Horace Silver, Bill Evans, Wes Montgomery, Jack Teagarden, Red Norvo, Stan Getz, and Nat "King" Cole. Silver's "Moon Rays" is an especially inspired choice. Although atypically lyrical, it shows off Silver's scrappy piano style and unpretentious approach to small-group orchestration. But with Evans already accounted for on Davis's "So What," the inclusion of the very similar "Blue in Green," by Evans's trio, seems unnecessary. And what is Montgomery doing here? Admittedly the most influential guitarist since Charlie Christian, he's an anomaly on a collection saluting musicians who have expanded jazz form, rather than those who achieved unparalleled virtuosity.

I wish that Williams had found space for Jabbo Smith, Ethel Waters, George Russell, Eric Dolphy, Albert Ayler, small group Lester Young, Billy Strayhorn's "Chelsea Bridge," something by the Miles Davis Quintet with Wayne Shorter and Tony Williams, even (dare I say it) "Giant Steps." I also wish Williams had included one of Cecil Taylor's recent solo piano performances to complement his "Enter Evening," recorded with a septet in 1966— Taylor is his generation's Art Tatum, as well as its Duke Ellington.

But what really disturbs me is the scant notice Williams pays to the last fifteen years—roughly one-fifth of jazz history, after all, and the period most in need of disquisition. In the booklet that accompanies *Classic Jazz*, he credits the World Saxophone Quarter with reaping "the fruits of two decades of modal and free jazz. Perhaps the group offers, on a small scale, the same kind of . . . synthesis of what . . . preceded them as Morton, Ellington, and Monk did before them." But he doesn't address why this synthesis supposedly took so long to occur. The answer, of course, is that it didn't. By the time the WSQ assembled in the late seventies, the Art Ensemble of Chicago and Air—AACM co-ops conspicuously absent from *Classic Jazz*—had already synthesized Coltrane's modal exotica and Coleman's collective improvisation, offering in the bargain a renewal of the spontaneous interaction between soloist and rhythm section that Williams admires in Mingus's ensembles and Max Roach's early work with Parker. "Major and influential events in jazz history center around its composers, who give the music synthesis and overall form, and around individual innovative soloists, who periodically renew its musical vocabulary," Williams writes, and for a jazz critic, these are words to live by. So why does Williams fail to acknowledge the contributions of Muhal Richard Abrams, John Carter, Anthony Davis, and Henry Threadgill, the composers now following the pattern of Morton, Ellington, and Monk in forging an orchestral syntax for what used to be called free jazz?

Wittingly or not, the revised *Classic Jazz* confirms the erroneous, widespread prejudice that jazz has reached the condition of epilogue, that everything new sounds like something we've heard before or is something so aberrant it's not worth hearing. Williams is too vigilant an observer to believe this, but despite his advocacy of the WSQ, he seems uninterested in contemporary developments. This troubles me, because if time has caught up with the most perceptive critic ever to write about jazz (one of Ornette Coleman's earliest supporters), what hope does the future hold for the rest of us?

Am I saying that Williams was wrong to tamper with *Classic Jazz?* Hardly. The revised edition may be an addendum rather than the update one hoped for, but its programming is superior to the original, and not the least of its virtues is its ability to provoke arguments while delivering bliss. Morton's "Black Bottom Stomp" and "Dead Man's Blues" are still here, and so are Armstrong's "West End Blues," "Struttin' with Some Barbeque," and "Sweethearts on Parade," Duke's "Ko-Ko" as well as Bird's completely different "Ko Ko," Monk's "Misterioso" and "Criss-Cross," Sonny Rollins's "Blue 7," Lester's penetrating honks on "Taxi War Dance," and an excerpt from Ornette's "Free Jazz." These are works so rich they reveal new detail on every hearing, especially in tandem with Williams's incisive running commentary. To ponder *Classic Jazz's* shortcomings is like struggling with some barbeque, when the thing to do is strut. Perhaps critical debate is something that should be kept in the family, and perhaps critics should remember that readers regard them as consumer advocates, not sages. All you really want to know is if *The Smithsonian Collection of*

*Classic Jazz** is worth the asking price. Yes, many times over. Just don't accept it as the last word.

(*July, 1987*)

JAZZ REP

I

The highlight—make that the vindication—of the American Jazz Orchestra's May 12, 1986, unveiling at Cooper Union came just before intermission. The easy-does-it John Lewis relinquished the baton to the Bernstein-like guest conductor Maurice Peress, who exhorted the AJO through a reading of Duke Ellington's *Harlem* that was tentative in places, impetuous in others, but finally thrilling in its headlong momentum. Like the earlier *Black, Brown, and Beige*, *Harlem* is paradigmatic Ellington: a mural of uptown life intended for display in the priciest midtown galleries. A twelve-minute tone poem allowing no improvisation but skillfully exploiting the unique tonal characteristics of each Ellingtonite to maximum advantage, *Harlem* was written aboard the *Île de France* on commission from Arturo Toscanini and the NBC Symphony, which apparently never performed it. There were actually three versions of the piece: the rarely heard symphonic version that Ellington prepared for Toscanini (entrusting the orchestration to Luther Henderson); the one Duke recorded in 1961, with just his big band; and one for both big band and symphony, which amounted to a compromise between the first two. The "jazz band" orchestration of *Harlem*—not only the most potent of the three, but arguably the most fully realized of Ellington's longer works—was an inspired choice as the centerpiece for the AJO's debut concert. Peress brought it back to life in a manner that sidestepped comparison to the original. The AJO's lusty performance was a triumph for jazz repertory, and conclusive evidence that a great jazz composer needn't be physically present for one of his works to receive a

---

* Smithsonian Collection of Recordings, Washington, D.C. 20560.

warm-blooded reading—not even a composer notorious for frequently eschewing notation.

Which, I assume, is precisely what Gary Giddins, the jazz critic for the
Village Voice, had in mind, when he convened the AJO. But for some of us in
general sympathy with Giddins's aims, the rest of the evening gave pause. A
week before the event, Giddins revealed in his Weather Bird column that he
envisioned a "neutral" orchestra, "not contingent on the compositions or biases
of a star leader." In other words, a jazz repertory orchestra along the lines of a
philharmonic—not a ghost band like Basie's and Ellington's have become, or a
larger version of any of the increasing number of small groups dedicated to the
works of a single composer (Dameronia, for example, or Mingus Dynasty).

Although I applaud John Lewis's appointment as the AJO's conductor, I
have to wonder whether his bias toward riff-based charts from the libraries of
Ellington, Count Basie, Jimmy Lunceford, and Fletcher Henderson compro-
mises that wished-for neutrality. Lewis's preference for such material is
understandable, given his apotheosis of the riff in his writing for the Modern
Jazz Quartet. But it's folly to expect an orchestra assembling in public for the
first time, and planning to regroup only a few times a year, to attain the group
chemistry that gave the riffing swing bands their boundless vitality. The AJO's
section work lacked bristle, and the drummer Charli Persip's tempos were a
fraction off all night, particularly on Sy Oliver's "For Dancers Only," where
one missed Jimmy Crawford's inimitable terpsichore from the original. Still,
these re-creations at least offered the pleasure of anachronism, an opportunity
to hear, in the flesh, "Jack the Bear," "Every Tub," "Lunceford Special," and
"King Porter Stomp" presented more or less as we know them from records. No
such justification was possible for two newly commissioned arrangements by
Slide Hampton, which failed in attempting to update Dizzy Gillespie's big
band. Hampton's enlargement of Charlie Parker's "Confirmation" was marred
by flaccid and historically incongruous clarinet and flute voicings, and his
Fantasy on 'Shaw Nuff' and 'Anthropology' gave rise to the spectacle of a
winded Gillespie (an unannounced guest) backing off from a showdown with
his taunting echo, Jon Faddis.

As members of the jazz press congregated at a reception after the concert,
some grumbled over Giddins and Lewis's apparent decision to have the
orchestra's soloists parrot recordings instead of going for themselves. To the
contrary, I think the re-created solos were the evening's saving grace (along with
Harlem). Every writer I know has, at some point, copied down the sentences of
Hemingway, James, and Faulkner just for the thrill of feeling the syntax in his
or her own knuckles. Loren Schoenberg approached Lester Young with the
same gleeful awe. The trumpeter Virgil Jones was in touch with both Bobby

Stark's melancholy and Red Allen's deadpan on the Henderson numbers, and his evocation of Cootie Williams on the most indelible of Ellington's concertos was uncanny. Still, the concert could have used more surprises like the one that the baritone saxophonist Hamiet Bluiett pulled on "Jive at Five." He treated the Basie original as a palimpsest, filling in the blanks with raunchy, interval-vaulting squeals that transported the chart into a contemporary realm without putting the period flavor irretrievably out of reach. Lewis was wise in awarding him two additional impromptu choruses (though there was grumbling over this, too, among the more conservative critics). But if one came to Cooper Union wondering what Craig Harris would sound like playing Ellington, or what Ted Curson would sound like bouncing off Basie, one left the hall still wondering. The majority of the solos went to studio vets and capable sight-readers Eddie Bert, Bob Milikin, and Walt Levinsky, while Harris, Curson, and Jimmy Heath (ace soloists with little recent big band experience) were mostly relegated to section work—a curious turnabout that, in most instances, resulted in wayward ensembles and workmanlike solos.

The American Jazz Orchestra was a good idea going in, and it remains a good idea, though a rethinking of priorities seems in order, with the success of *Harlem* showing the way, one hopes, to an emphasis on composition that would make the goal the workable one of reinterpreting (for example) Duke Ellington's written scores, not the hopeless one of replicating his orchestra and soloists. Think of such masterpieces and near-misses as Ellington's "Reminiscing in Tempo" and "The Three Black Kings," Mingus's "Half Mast Inhibition" and "Don't Be Afraid, the Clown's Afraid, Too," George Russell's "All About Rosie," and "Cubana Be" and "Cubana Bop," Ferde Grofé's "Metropolis," Ralph Burns's "Summer Sequence," Robert Grattinger's "City of Glass," and Lewis's own "Three Little Feelings," to say nothing of the masterpieces and near-misses to come if the AJO makes good on its promise to commission new works. Then you have an inkling of what this repertory orchestra's proper repertory should be.

(*May, 1986*)

II

Last December, a few months after Benny Carter's eightieth birthday, the National Academy of Recording Arts and Sciences presented him with a Lifetime Achievement Grammy. Nothing that the hostess Dionne Warwick read in introducing Carter to the network television audience explained why he was being honored. She delivered generic blather about Duke Ellington and Cab Calloway, Harlem and the Cotton Club. Then Carmen McRae sang "Body and Soul," a standard linked to Coleman Hawkins, and David Sanborn and Carter duetted on "Just Friends," a number associated with Charlie Parker. It was as though Carter was being honored simply for living to a ripe old age, and because NARAS was eager to present an elderly jazz great—*any* elderly jazz great, damn it—with what amounted to an equal-opportunity Grammy for prestige and roots musics that don't move as many units as rock (the other honorees included Roy Acuff, Isaac Stern, and B. B. King).

The show's writers should have provided Warwick with copy that mentioned Carter's important role in codifying big band jazz (as an arranger for the Fletcher Henderson Orchestra and *de facto* music director for the Chocolate Dandies, McKinney's Cotton Pickers, and the Wilberforce Collegians), his savvy as a talent scout (although his own big bands never caught on with the public, they spawned Miles Davis, J. J. Johnson, Art Pepper, and Max Roach), his rank as one of the two most influential alto saxophonists before Charlie Parker (the other was Johnny Hodges), his skills as a multi-instrumentalist (he also plays Armstrong-style trumpet, and has recorded on piano and trombone), and his pioneering efforts on behalf of racial integration (he was the first black composer to penetrate Hollywood, and one of the first black bandleaders to employ white sidemen during an era when integration still meant blacks working under a white boss). A big band should have been assembled to play Carter's "When Lights Are Low" (his most frequently performed composition, from 1945), "Symphony in Riffs" (supposedly the first big-band arrangement to score the saxophones in block chords, from 1933), "Waltzing the Blues" (the first jazz tune in $\frac{3}{4}$ time, from 1937), or—because sales are finally all that matter to NARAS—"Cow Cow Boogie" (a 1942 hit by Freddie Slack and Ella Mae Morse, co-written by Carter, Don Raye, and Gene DePaul).

Carter was paid a more meaningful tribute in February 1987, when he performed an SRO concert with the American Jazz Orchestra at Cooper Union in New York. *Central City Sketches* (Musicmasters CIJD-20126Z/27X), Carter's first big-band album in two decades, recorded with the AJO a week after the concert, suggests that not the least of his abilities is hoodwinking Father Time. His blues choruses on "Easy Money," to cite one example among many, confirm that he remains one of jazz's most dazzling improvisers, still in full possession of the roseate, almost Marcel Mule–like tone that has been his signature for over half a century, but thoroughly modern in his harmonic values and rhythmic placements. The program offers a Carter retrospective, with material ranging from "Lonesome Nights," "When Lights Are Low," the aptly titled "Symphony in Riffs," and "Blues in My Heart" (unaccountably

burdened with a doubled-up and dated-sounding "contemporary" beat—
Carter's only injudicious revision) from the thirties, to the ingratiatingly varied
six-part title suite presented as a work-in-progress at the New York concert and
finished just in time for the recording session (the melodic accelerations of the
section subtitled "Promenade" are especially winning). What makes such a
retrospective invaluable is that Carter seldom has an opportunity to perform his
earlier pieces anymore, because the younger musicians he generally plays with
don't know them. (They *do* know "When Lights Are Low," but only because
Miles recorded it. Otherwise, Carter is in the uncomfortable position of having
outlived his repertoire.) The set also marks an auspicious recording debut for
the American Jazz Orchestra, with music director John Lewis spelling Dick
Katz at the piano for sharp-witted choruses on four numbers, including Carter's
no-doze arrangement of the Fred Waring warhorse "Sleep." The AJO was
formed to show that what's needed to keep the mold off the classic jazz is the
fresh air of performance. *Central City Sketches* proves the point and then some.

(February/April, 1988)

III

Though the 1988 JVC Festival all but ignored new directions, it fulfilled the
equally vital mission of resurrecting great works by dead composers. At
Carnegie Hall on June 26, Maurice Peress conducted the American Composer's Orchestra (augmented by such guest soloists as Jimmy Heath and Sir
Roland Hanna) performing four of Duke Ellington's extended compositions.
The following night, an eleven-piece ensemble made up, for the most part, of
former Charles Mingus sidemen or current members of Mingus Dynasty,
tackled Jimmy Knepper and Sy Johnson's arrangements of Mingus compositions at the 92nd Street Y.

The Ellington concert had the aura of an Event: the first all-Ellington
symphonic concert ever, according to the Playbill. (It was also a homecoming
of sorts: Benny Goodman may have brought jazz to Carnegie Hall in 1938,
but it was a 1943 Ellington concert that proved that jazz belonged there.) The
program opened with Luther Henderson's symphonic orchestration of *Les
Trois Rois Noirs (The Three Black Kings)*, a three-part ballet suite commissioned
by the Dance Theatre of Harlem sometime in the early seventies, but not given
its premiere until 1977, when Mercer Ellington added the finishing touches. In
order, the work's three movements are dedicated to King Solomon, Balthazar
(the black king of the Magi), and Martin Luther King. The first movement—
with its opening piano and percussion fandangos, its skirmishes with atonality,

its bombast and bustle and contrasting interludes of tranquillity—is major Ellington, and the ACO gave it a spirited interpretation. But Heath's spotlight tenor saxophone improvisation on the second movement was more beholden to bebop than to Ellington, and the concluding movement—a churchy waltz featuring Heath's soprano above prim strings—was contrived and anticlimactic, however catchy. (Do we have Mercer to blame for this ending?)

Following *Les Trois Rois Noirs*, the orchestra interpreted the first three movements ("Work Song," "Come Sunday," and "Light") from *Black, Brown, and Beige*, the landmark seven-part "tone parallel to the history of the American Negro" that Ellington wrote for his 1943 Carnegie Hall concert. Despite numerous felicities—expecially Frank Wess's perfect evocation of Johnny Hodges on "Come Sunday"—*BB&B* showed stretch marks in this symphonic version prepared by Peress at the composer's request in 1969. But the concert reached a crescendo with Peress's symphonic orchestration (from Ellington's original big-band score) of *New World A-Comin'*, and sustained it with Peress's revision of Henderson's original symphonic treatment of Ellington's *Harlem*. *New World A-Comin'*, a piano concerto that Ellington first performed at his second Carnegie Hall concert, in 1944, glistened in this performance. Hanna, who shares Ellington's impressionistic bent and affection for ragtime, was an ideal choice to interpret Peress's transcription of the composer's original piano solo. Investing the performance with such a combination of amusement and passion that only the presence of someone to turn the pages for him revealed that he wasn't playing off the top of his head, Hanna showed that there's more to knowing the score than reading notes off paper.

On *Harlem*, the star was the conductor, who was literally in the air, with both feet off the ground, at the climax. The ACO—more accustomed to performing such contemporary classicists as John Harbison and Ellen Taaffe Zwilich—gave him precisely what he wanted: even the strings affected an Ellingtonian growl. Peress has become something of a crusader for Ellington, just as Leonard Bernstein (for whom Peress understudied with the New York Philharmonic) used to be for Mahler and Ives, and he's sophisticated enough in his appreciation of idiomatic American music to know that the concert hall needs Ellington more than Ellington needs it. But because the concert was presented under the aegis of JVC Jazz, and not as part of an orchestra's regular season, it represented only a partial victory for the Symphonic Ellington. Even so, the evening went a long way toward countering Ellington biographer James Lincoln Collier's arrant nonsense about Ellington's reach exceeding his grasp in the area of extended composition.

The Mingus Big Band shared its bill with the pianist Geri Allen and her trio, who played a meandering set of Allen originals. The Mingus portion of the program got off to a promising start with an animated version of "Jump Monk," featuring solos by every member of the band except for the drummer Billy Hart. Everyone was in good form, with the trombonist Knepper, the trumpeter Jon Faddis, the pianist Jaki Byard, and the tenor saxophonist David Murray copping top honors. The concert took a bad turn, though, when Knepper, who doubled as conductor (thus depriving the group of one of its most expressive voices at strategic moments), counted off "The Shoes of the Fisherman's Wife," a quicksilver Mingus tone poem. The brass and reeds were sluggish in responding to one another, and the rhythm section sounded uncomfortable with Mingus's precipitous time shifts and tempo changes.

The news that this band would be performing the first three movements (roughly the first half) of *The Black Saint and the Sinner Lady*, Mingus's 1963 album-length *cri de coeur*, was what lured many of us to Ninety-second Street. But in proportion to its ambition, *The Black Saint* was the evening's cruelest disappointment. The alto saxophonist John Handy tried gallantly but failed to approximate the "tears of sound" that Mingus, in his liner notes to the recording, credited Charlie Mariano for achieving. And the work's Ellingtonian lineage was obscured, largely because the baritone saxophonist Nick Brignola didn't plumb his horn's bottom, as Jerome Richardson did on the Mingus album, paying homage to the Ellington Orchestra's Harry Carney. (In recognizing Ellington as both inspiration and competition, Mingus was Norman Mailer to Ellington's Ernest Hemingway.) Nor was Craig Handy a suitable replacement for the versatile journeyman Richardson on soprano saxophone. Handy wasn't necessarily wrong for taking a different approach from Richardson's, but his solo slowed down the tempo instead of accelerating it—a grave miscalculation from which the band never recovered.

The performance's most agonizing flaw was its lack of Mingus-like vehemence. Unlike Mingus's groups, this ensemble never sounded like it was getting something off its chest. Nobody sounded mad at anybody else in the band, and no one seemed bent on facing down the audience, the way that Mingus always did. As a result, this wasn't Mingus. At one point, Knepper asked the stage crew to bring the lights up. "After all, we're trying to read up here," he said—which was exactly the problem. It *sounded* like the musicians were reading. They should have had the piece memorized (and maybe they would have, if they had been allowed more rehearsal time). In any event, it was painful to watch the great Jaki Byard tentatively reading passages that he probably made up from scratch (at the composer's angry urging) twenty-five years earlier.

Still, there's no evidence that Mingus ever performed *The Black Saint and the Sinner Lady* in concert, and nobody knows how many splices there were on the 1963 recording. I hope that this band gets another crack at this epochal piece, and I want to be there when it does, just as I do the next time an orchestra plays *Les Trois Rois Noir.*

(July, 1988)

IV

Jazz people are no different from most folks in wanting what they can't readily have. Fifty years ago, when jazz meant big bands to most Americans, musicians and knowledgeable listeners longed for the informality of small groups. Now that big bands are in danger of extinction, the sight and sound of one just tuning up is enough to induce yearning for a time when jazz enjoyed strength in numbers in the audience as well as on the bandstand. The decline in popularity of the big bands after World War II was symptomatic of a dwindling interest in jazz in general as it evolved into bop. A desire for big bands isn't necessarily the same thing as nostalgia for the sounds of the big-band era, or else why would so many musicians associated with the jazz avant-garde be forming large ensembles? Big bands have something to offer everyone: for the composer/arranger, a wider spread of colors; for the instrumentalist, a chance to compare notes with section mates; for the listener, the overwhelming physical vibration of hearing as many as seventeen voices blending into one—a thrill that's become all the more seductive now that we can no longer take it for granted.

By now, you're probably expecting me to announce that we're in the midst of an unaccountable big-band revival, but that would be wishful thinking. Though a surprising number of this year's outstanding releases (by leaders as disparate as Julius Hemphill, Bill Holman, and Chris McGregor) feature big bands, most of these outfits assemble only for record dates and infrequent club or festival appearances. A happy exception is the powerhouse that tenor saxophonist Illinois Jacquet leads on *Jacquet's Got It!* (Atlantic Jazz 7 81816-1), a release that actually warrants the exclamation point after its title.

What Jacquet's got, at long last, is the big band he needed to certify his stardom after his hit record "Flying Home" with Lionel Hampton in 1942. Jacquet is the archetypal tough tenor—a real bruiser on up-tempos who mitigates swagger with just enough manly sentimentality on ballads. (Tough tenors are like the fictional hard-boiled private dicks: women turn them to mush.) Like all tough tenors, Jacquet's a descendant of the late Herschel Evans, Lester Young's fuller-toned alter ego in Count Basie's first big band. So it's a pleasurable jolt to hear Jacquet evoke Evans rather than Young on *Jacquet's*

Got It!'s version of "Tickle Toe," because the new arrangement (credited to Jacquet, Wild Bill Davis, and Carol Scherick) is otherwise so faithful to the Basie-Young 1940 original. In Phil Wilson, Eddie Barefield, and Davis (whose "April in Paris" for Basie was no fluke), Jacquet has arrangers who know how to use other horns to add ripple to his muscle. As a result, he sounds majestic, especially on Wilson's plush arrangement of "More Than You Know" and Davis and Scherick's adaptation of "You Left Me Alone," an obscure ballad by Jacquet and the late Tadd Dameron, the most melodically inclined of the bebop composers. But the knockout cut is Jacquet's own "Blues from Louisiana," with a vibrant stop-time solo full of long, plunging low notes that sounded blue in more ways than one.

In addition to its more obvious virtues, Jacquet's band benefits from an ideal mix of experience (the trumpeter Irv Stokes, the bassist Milt Hinton, and the saxophonists Barefield, Marshall Royal, and Rudy Rutherford) and youth (the trumpeter Jon Faddis, the trombonist Frank Lacy, and the alto saxophonist Joey Caveseno). With slightly younger and less star-studded personnel, Jacquet last month played the first four nights of a two-week-long Big Band Festival at the Blue Note in Greenwich Village. I caught him on his closing night, and his band was even more rewarding live than on record, thanks in no small part to his un-self-conscious showmanship (he even sang "Don't Blame Me" and played alto, glissing like Johnny Hodges) and his valuable knack for conducting with his shoulders, elbows, hips, knees, and horn. Kenny Bolds, who over-played his rim shots behind Jacquet's solos, was no match for Duffy Jackson, the drummer on the record (and Mel Lewis's heir apparent as the best big-band drummer in the business), in terms of efficiency or relaxed propulsion. Even so, he drove the band hard, which is finally all that really matters. This wasn't the most fun I've ever had while hearing Illinois Jacquet—that had to be the night about seven years ago when I chatted with Norman Mailer on line for the men's room at Sweet Basil (he was kind to me, even when I drunkenly exclaimed, "You really do affect an Irish brogue!")—but it was close.

At the Blue Note, Jacquet sported white linen and his sidemen wore fire-engine-red blazers. In contrast, the members of the American Jazz Orchestra—the next band at the festival—wore dark blue, and their versions of Duke Ellington and Billy Strayhorn were also muted and self-effacing. This repertory ensemble has been in residence at Cooper Union since 1986, and it's probably the only big band ever fronted by a jazz critic who wasn't also an instrumentalist or composer. Although John Lewis conducted, Gary Giddins announced most of the tunes and at times seemed unaware that he wasn't still preaching to the unconverted among the subscribers at Cooper Union. What Giddins had to say about Ellington was unimpeachable, but the Friday night overflow crowd I was a part of probably didn't need this crash course. It risked making the AJO's performance the artistic equivalent of a good deed (the PBS approach to pleasure), and maybe this was why the audience was so circumspect, especially compared to the one for Jacquet the night before.

This may have been just as well, because it enabled one to concentrate on the wealth of small detail in the AJO's Ellington interpretations. That the AJO chose an all-Ellington and Strayhorn program for its nightclub debut was fitting, because it was Ellington's death and the realization that jazz's most protean composer would no longer be around to play his own works that lent urgency to the jazz repertory movement. Although Mercer Ellington continues to tour with his father's orchestra, what audiences usually demand of him is medleys of Ellington's greatest hits, so the AJO is performing a valuable service in reanimating neglected masterpieces like Ellington's "Sepia Panorama" and "Ko Ko," and Strayhorn's "Johnny Come Lately." In scaling Ellington down to size, small groups frequently uncomplicate him, ignoring, for example, the slightly dissonant countermelodies that gave his ballads backbone. To their credit, Lewis and the AJO put the complications back in, so that even the more familiar pieces become fraught with surprise. And to actually *see* with which combinations of instruments Ellington achieved some of those mysterious voicings is an education that records can't offer.

The highlight of the set I heard was the famous saxophone chorus on "Cottontail," which I don't think I ever before realized was the tacit inspiration for Horace Silver's "Sister Sadie" and countless other soul-jazz compositions from the late fifties and early sixties. The tenor saxophonist Loren Schoenberg did an impressive job of bringing Ben Webster back to life on this piece; Norris Turney (an actual Ellingtonian) made Johnny Hodges's alto swoops his own on "Warm Valley"; and the pianist Dick Katz wryly approximated the maestro's dissonances, offbeats, and ripples. The two most irrefutable justifications for jazz repertory were on view at the Blue Note. One was the sight of younger musicians, like Schoenberg and the excellent bassist John Goldsby, capable of playing older styles with élan. The other was the sight of big-band veterans like Turney, the trombonist Eddie Bert, and the baritone saxophonist Danny Bank as living reminders that we're not talking about antiquity when we talk about classic jazz—we're talking about the recent past. For all of that, the set lacked some unnameable something. Though the band swung (how could it not, with Mel Lewis on drums?), it did so a shade too politely, and the music often sounded flat—not harmonically, but emotionally.

I confess that Giddins's involvement with the AJO fascinates me, as I assume it does most of our fellow critics. The potential for conflict of interest (and for the *appearance* of it, which can be just as damaging) is, of course, enormous, and Giddins hasn't always been scrupulous about avoiding it: in his last Weather Bird column before taking a year's leave of absence from *The Village Voice* in 1987, he praised new releases by Lewis and the AJO trombonist Eddie Bert (both albums deserved the praise, but that's hardly the point). Conflict of interest is only the half of it, though. The real danger for a critic in becoming too directly involved in the making of jazz—as opposed to examining the result—is that he'll come to overvalue chops, sight-reading skills, and other superficial aspects of "musicianship," just as many musicians do (by that standard, some original Ellington and Basie sidemen wouldn't be good enough for the AJO).

But the AJO perplexes me beyond Giddins's involvement, and so does Jazz Rep as an eighties phenomenon. In *The Worlds of Jazz* (Grove Press, 1972), one of the most peculiar books ever written about music or anything else, André Hodeir spins a prophetic fable about the Aldebarron brothers, Fenimore Winthrop and Bartholomew Cleophus (nicknamed "Doc," naturally), two latecomers to jazz scholarship who, deciding that the music lacks a repertoire, form an orchestra to address the need. "What interested them was not the mass of themes drawn on by jazz bands large and small, but very specifically the arrangements and original compositions which were written for them— preferably at a period remote enough to be regarded as historical—and which were no longer played," writes Hodeir (as translated by Noel Burch). There was just one problem: "Confronted with an old score, young musicians would of their own accord recreate the interpretive style of the period, said Doc; they would find a new style, said Fenimore Winthrop." Here we have the central dilemma of Jazz Rep. Neither approach is necessarily better than the other, but confusion results when you try to have it both ways, as the AJO and other reportory ensembles frequently do.

The other problem is that after a rocky start, the AJO has become a more cohesive ensemble only by narrowing its horizons. In focusing on Ellington and swing, the AJO has strayed from its original mission of performing classic, neglected, and newly commissioned works in all jazz styles. But what single orchestra could have accomplished so utopian a goal? I'm asking too much of Giddins's brainchild. We need another orchestra like the AJO, but one specializing in bop and post-bop idioms, and still another (along the lines of the American Composers Orchestra or the defunct Jazz Composer's Orchestra) to commission new works. In order for one ensemble to do all that, it would need a floating personnel and municipal sponsorship similar to that routinely lavished on philharmonics. This isn't going to happen soon (or possibly ever), but in the meantime we need more big bands of every conceivable kind. Jazz just wouldn't be the same without them.

(December, 1988)

AT THE MOVIES

EVERYCAT

In jazz circles, the early word on *'Round Midnight*, the French director Bertrand Tavernier's nicotine-stained valentine to bebop in European exile in the late fifties, went roughly as follows: critics would loathe the movie for its trivialization of jazz history, but musicians—flattered to see one of their own on the big screen—would adore it for validating their existence (the "I'm in Technicolor, therefore I am" impulse that made longhairs embrace *Easy Rider* in the late sixties, and black urban audiences embrace *Shaft* and *Superfly* a few years later). Musicians for, critics against, is indeed the way the sides are lining up, now that *'Round Midnight* has opened. You can probably guess which side I'm on, but I'm not saying that musicians are wrong.

'Round Midnight—starring the tenor saxophonist Dexter Gordon as Dale Turner, a fictional composite of Lester Young and Bud Powell—is about jazz as a religious experience, with all the stigmata and stations of the cross presented in jumbled, vaguely sacrilegious fashion. Gordon's Dale Turner is a tortured black innovator who, like Young, memorizes the lyrics to songs before interpreting them instrumentally, addresses even male acquaintances as "Lady," and spent time in the stockade during World War II for carrying a photograph of his white wife. Like Bud Powell, Turner was once beaten repeatedly on the head with billy clubs, and like many musicians of Powell's generation, he is easy prey for obsequious drug pushers and sleazeball promoters (typified here by Martin Scorsese, in a distracting cameo). He has an old sidekick nicknamed Hersch (presumably Herschel Evans, Young's sparring partner in the Count Basie Orchestra), a daughter named Chan (after Chan Richardson, Charlie Parker's common-law wife), a lady friend called Buttercup (just like Powell's widow), and another who sings with a white gardenia pinned

in her hair (just like you-know-who, though the buppie princess Lonette McKee is unlikely to remind you of Billie Holiday). When the man standing next to Dale at the bar passes out, Dale says "I'll have what he's been drinking," just as legend has it Young once did. And like Young, he calls someone shorter than himself "half-a-motherfucker"; the only problem is that Lester was talking to Pee-wee Marquette, the midget master of ceremonies at Birdland, whereas Dale is addressing the normal-sized Bobby Hutcherson.

You get the point: Turner is Everycat, less a character than an accumulation of fact and lore. Despite this, 'Round Midnight, in its meandering middle stretches, is less a jazz film than another buddy-buddy flick, replete with unacknowledged homoerotic undertones (one scene in which Turner is writing music at the opposite end of the table from his French graphic-designer roommate and benefactor—played by François Cluzet and faithfully modeled on Powell's keeper, Francis Paudras—plays like an inadvertent parody of the successful two-career marriage). The only difference is that one of the buddies is a black, dypsomaniacal, six-foot-seven *down beat* Hall-of-Famer.

Even so, it's easy to understand why musicians are pleased with 'Round Midnight. Clichés and all, it's as sympathetic an account of the jazz life as has ever been presented in a feature film, erring on the side of compassion rather than exploitation, guilty of sentimentality but not sensationalism. The uncertainty of Gordon's line readings betrays that he's no actor and that he was given no real character to work with. But his presence and dignity—his paunch-first stagger, his big-man's daintiness, his rasped expletives, and his vanquished Clark Gable good looks—rescue the movie from banality. A former alcoholic, drug abuser, and longtime expatriate himself, he's obviously drawn from personal experience to give a performance that one suspects would have been beyond the ability of a more experienced actor. His peers will recognize themselves in him, and they can be proud of what they see.

Oddly, the drawback to casting Gordon in the lead role was musical. When he's in peak form, Gordon's tone is as bracing and aromatic as freshly perked coffee. But he was recovering from assorted illnesses and an extended period of inactivity during filming, and as a result, his solos have a spent, desultory air. In dramatic terms, this may be just as well, inasmuch as we are given to understand that Dale Turner is a man slowly snuffing himself out, capable of summoning up his former brilliance only in flashes, convinced that death is nature's way of telling him to take five. (You wonder what Francis is using for ears when he says that Turner is playing "like a god."*) But a sub-par Gordon

---

* You also wonder what Tavernier is using for ears, because, contrary to what our own ears tell us, Francis is supposed to be telling it like it is. It's worth passing along an astute comment that the pop critic Ken Tucker made about this sketchily drawn character, after the screening we

makes the soundtrack album (Columbia SC-40464) pretty tough going. Gordon isn't the only culprit; the soundtrack's supporting cast is made up of musicians ten to twenty years his junior, for whom bebop is little more than a formal exercise, and Herbie Hancock's incidental music is flat and uninvolving when divorced from the film's imagery. Gordon deserves the plaudits he's winning as an actor, but it would be a pity if the lay audience now discovering him accepts the music from '*Round Midnight* as characteristic.

Although '*Round Midnight* is the only recent movie to star a jazz musician, it's not the only one with a jazz soundtrack. Spike Lee's sleeper hit *She's Gotta Have It* boasts a fine soundtrack by his father, the bassist Bill Lee, which has just been released on Island 7 90528-1. The elder Lee's modest, by turns moody and frolicsome small-band score goes awry only once, exactly where the black-and-white movie does: in the too-sweet Ronnie Dyson vocal accompanying an oversaturated Technicolor ballet. But in its mix of disciplined composition and footloose improvisation, Lee's music recalls earlier film scores by such jazz composers as Duke Ellington (*Anatomy of a Murder*), Miles Davis (*Frantic!*), Sonny Rollins (*Alfie*), John Lewis (*Odds Against Tomorrow*) and Gato Barbieri (*Last Tango in Paris*). It also brings to mind Henry Pleasants's conjecture that the collaborations between composers and film directors have the potential to become the modern equivalent of lyric theater. Writing before the corporate takeovers of both film and record companies, and before the success of *Easy Rider*, *The Graduate*, and *Saturday Night Fever* made soundtrack albums little more than K-Tel greatest-hits collections in disguise, Pleasants had no way of knowing that film composers would eventually rank lower than the music-acquisitions lawyers in the overall scheme of things. It's becoming more and more unusual to hear a score like Bill Lee's, brashly original and homogenetic to the film it serenades. If Dexter Gordon's haunting portrayal in '*Round Midnight* suggests that jazz musicians can be riveting onscreen subjects, Lee's score confirms that they also have plenty to offer behind the scenes. Here's hoping that more film producers take them up on the offer.

(*November, 1986*)

---

both saw. Francis is a commercial artist; we see one of his posters for an American film starring Jeff Chandler. Tucker pointed out that such a Frenchman would almost certainly be obsessed with American popular culture in general, not just jazz. But when Francis accompanies Dale to New York, he doesn't go looking for paperbacks by Erskine Caldwell or movies by Nicholas Ray. For that matter, he doesn't even hear any live jazz, except for Dale's.

BIRDLAND, MON AMOR

Charlie Parker assured himself of immortality when he recorded "Ko Ko" for Savoy Records, on November 26, 1945. This wasn't the first time that bebop was performed in a recording studio, nor was "Ko Ko" the first jazz "original" extrapolated from the chord sequences of an existing tune—a practice that didn't begin with bop, contrary to popular belief. For that matter, Parker wasn't the first improviser to recognize that despite the nondescript melody of Ray Noble's "Cherokee," its fast-moving chords held the potential for tour de force; Charlie Barnet had beaten him to it by six years. Yet there was really only one historical precedent for "Ko Ko": Louis Armstrong's 1928 recording of "West End Blues." As Armstrong had done (and as John Coltrane would later, with "Chasin' the Trane"), Parker with one performance reshaped jazz into his own image by establishing an exacting new standard of virtuosity. Listeners encountering "Ko Ko" for the first time are likely to be astonished by Parker's faultless execution at a tempo that starts off reckless and seems to speed up as it goes along. But the most remarkable aspect of "Ko Ko" is Parker's reconciliation of spontaneity and form—the impression of economy despite the splatter of notes; the surprising continuity of suspenseful intro, staccato bursts, pulsating rests, and phrases so lengthy they double back on themselves at the bar lines. Parker's contemporaries faced the challenge not only of matching his technique, but also of emulating his harmonic and rhythmic sophistication, and his successors still face the same challenge.

Parker's innovations—and those of Dizzy Gillespie and Bud Powell—are today so ingrained in jazz that it's difficult to remember that bebop was initially considered so esoteric and forbidding that only its originators could play it. "Ko Ko" would seem to prove the point. Stimulated by Parker, Max Roach made a breakthrough of his own on "Ko Ko," with an unyielding polyrhythmic accompaniment that amounted to a second melodic line. But Gillespie, who had been forced into service as a pianist in relief of Sadik Hakim (listen to Hakim's disoriented intro on "Thrivin' on a Riff," recorded earlier at the session, and you'll know why), also had to spell Miles Davis on "Ko Ko." Davis, then still in his teens and making his recording debut, declined even to try to his luck on the piece.

Parker was twenty-five but already addicted to heroin; he would be dead in less than ten years. Two weeks after recording "Ko Ko," he travelled with Gillespie to Hollywood for a nightclub engagement that lasted almost two months, despite the generally hostile reaction of Southern California audiences to bebop (the new style had been nurtured in secret on the East Coast, its dissemination hindered by a musicians' union ban on new recordings and by wartime restrictions on materials needed to manufacture discs). Parker didn't return to New York with his bandmates; instead, he cashed in his airline ticket to buy drugs. He found himself stranded in Los Angeles during a time when police crackdowns on heroin sent street prices soaring and often made the drug available at any price. In August, 1946, Parker was confined at Camarillo State Hospital after being arrested for setting fire to his hotel room in the aftermath of a disastrous recording session.

Parker spent six months at Camarillo, returning to New York in April of 1947. He then began a period in which he could do no wrong, at least in the recording studio, where he produced an unbroken succession of masterpieces for Dial and Savoy, including his most memorable ballad performance, a harmonic tangent on George Gershwin's "Embraceable You." Already married and divorced twice, he wooed two women almost simultaneously, marrying one in 1948 and setting up house with the other two years later, without bothering to divorce the first. In 1949, he scored a triumph at the International Jazz Festival in Paris, and a year later made the first of several records on which he was accompanied by woodwinds and strings, the format that brought him his greatest popular success.

But he never kicked his drug habit for good, and he drank to such excess that his weight ballooned to more than two hundred pounds. Despite his drawing power, nightclub owners became increasingly reluctant to book him, for fear that he'd show up in no shape to perform, or not show up at all. At one point, he was banned from Birdland, the Broadway nightclub named in his honor in 1950. Although he somehow eluded arrest for possession (there were rumors that he pointed detectives to other users in order to save his own skin), the cabaret card he needed in order to perform in New York City nightclubs was taken from him without due process, at the recommendation of the narcotics squad, in 1951. The incident that is said to have finally broken him was the death from pneumonia of his two-year-old daughter by his common-law wife, Chan Richardson, in March, 1954. Later that year, he swallowed iodine in an unsuccessful suicide attempt. He died of lobar pneumonia on March 12, 1955, while watching television in the New York apartment of the Baroness Pannonica de Koenigswarter, a wealthy jazz patron. He was thirty-four, but physicians estimated his age at fifty to sixty.

Forest Whitaker plays Parker in *Bird*, a film produced and directed by Clint Eastwood and written by Joel Oliansky. Miming to Parker's actual solos, with his eyes wide open and his shoulders slightly hunched and flapping, Whitaker captures the look we recognize from Parker's photographs and the one surviving television kinescope of him (a 1952 appearance on Earl Wilson's *Stage*

Entrance, which is featured in the excellent jazz documentaries *The Last of the Blue Devils* and *Celebrating Bird: The Triumph of Charlie Parker*). Unfortunately, even though he's been outfitted with a gold cap over one incisor to make his smile shine like Parker's, Whitaker is less convincing offstage, where most of *Bird* takes place. On the basis of his brief but effective turn as the young, possibly psychotic pool shark who spooks the master hustler played by Paul Newman in Martin Scorsese's *The Color of Money,* Whitaker was the right choice to play Parker—a master con, among other things. But Whitaker's performance is too tense and pent-up to bring Parker to life, and by the end of the movie, the actor seems as much the victim of heavy-handed writing and direction as the character does.

Why is it always raining in jazz films, and why are the vices that kill musicians always presented as side effects of a terminal case of the blues? It merely drizzled throughout *'Round Midnight,* and though the movie was false in other ways, the mist was in keeping with the slow-motion music performed by the ailing Dexter Gordon. The music in *Bird* is supposed to be defiant and ebullient, but the *mise en scène* is downbeat, with rain gushing against the windows of melodramatically underlit interiors. (You come out of the theater squinting, just like Eastwood.) Like Milton's Lucifer, whither this Bird flies is hell. He brings rain and darkness with him wherever he goes. His unconscious is haunted by symbols—or, to be more specific about it, by a literal cymbal that flies across the screen and lands with a resounding thud every time he drifts off or nods out. The vision is based on a (perhaps apocryphal) incident to which Eastwood and Oliansky have given too much interpretive spin. As an untutored seventeen-year-old in Kansas City, Parker is supposed to have forgotten the chord changes to "I Got Rhythm" while playing at a jam session with the drummer Jo Jones, who, legend has it, threw one of his cymbals to the floor as a way of gonging the teenager off the stage. Except for overwrought conjecture by Ross Russell in a purple passage toward the end of *Bird Lives!,* there is nothing in the voluminous literature about Parker to suggest that this public humiliation haunted him for the rest of his life. To the contrary, it's usually cited as the incident that strengthened his resolve to become a virtuoso. But in *Bird's* retelling, the echo of that cymbal deprives Parker of all pleasure in his accomplishments. *Bird's* Parker wants to rage but can only snivel, even when hurling his horn through a control-room window in abject frustration. You don't believe for a second that this frightened sparrow could have summoned up the self-confidence to make a name for himself in the competitive world of jazz in the 1940s, much less set that world on its ear with "Ko Ko." Parker was a compulsive, which is another way of saying that he was a junkie, but he was also obsessive, which is another way of saying that he was an artist.* Parker's torment is here, but not his hedonism or his genius or the hint of any connection between them.

Much of what *Bird* tells us about Parker is hooey, and at least one of its inventions is an abomination—a character, a slightly older saxophonist who knew Parker as an upstart in Kansas City, who becomes jealous when he finds out that Parker is the talk of New York. A final encounter with this saxophonist seals Parker's doom. Parker stumbles down Fifty-second Street, dazed to find that the jazz clubs that were the settings for his early triumphs have given way to strip joints. (The excuse for his surprise is his having been holed up in the country with Chan for a few months, but anyone who knows anything about jazz during this period has to wonder if he's been on the moon—articles in the national press were bemoaning the departure of jazz from "The Street" as early as 1948, and this is supposed to be 1955.) Told by another acquaintance that he hasn't seen anything yet, Parker wanders into a theater where his old Kansas City rival is knocking 'em dead with a greasy rhythm 'n' blues à la King Curtis. This triggers Parker's final breakdown. Even assuming that it was necessary to invent a fictional nemesis for Parker, why name that character "Buster," which the filmmakers should have known was the name of one of Parker's real-life Kansas City mentors, the alto saxophonist Buster Smith? And why pretend that Parker, who reportedly found good in all kinds of music, would have been shocked into a fatal tailspin by the advent of rock 'n' roll?**

Jazz fans appalled by the fraudulent portrayal of Parker won't be the only moviegoers displeased with *Bird*. It's a mess. Even at the epic length of two hours and forty-five minutes, the narrative feels hurried and absentminded, with more flashbacks within flashbacks than any movie since Jacques Tourneur's 1947 *noir, Out of the Past*. You're never sure who's remembering what, what year it is, how famous Parker has become, or how long he has to live— Whitaker has the same puppy-dog look no matter how far gone he's supposed to

---

*Charlie Parker to Paul Desmond (as quoted by Stanley Crouch, in *The New Republic*, February 27, 1989): "I put in quite a bit of study into the horn, it's true. In fact, the neighbors threatened to ask my mother to move once when I was living in the West. They said I was driving them crazy with the horn. I used to put in at least eleven to fifteen hours a day. I did that for over a period of three or four years."

**Gigi Gryce on Charlie Parker (as quoted by Orrin Keepnews in *The View from Within*, Oxford University Press, 1988): "We might be walking along and pass someplace with a really terrible rock and roll band, for instance, and he'd stop and say 'Listen to what that bass player's doing,' when I could hardly even hear the bass."

be, so the only way of telling is by the hair style on Diane Venora, the actress who plays Chan Richardson (she gives up her bohemian bangs and braids after becoming a mother). In terms of explaining to an audience that knows nothing about jazz (most moviegoers, in other words) what made Parker's music so revolutionary, *Bird* is about as much help as *The Ten Commandments* was in explaining the foundations of Judeo-Christian law—you almost expect someone to point to Parker and proclaim, like Yul Brynner as Pharaoh, "His jazz *is* jazz." The script primes us for ironic payoffs it never delivers; as when, for example, Parker hears his blues "Parker's Mood" sung by King Pleasure, whose lyrics envision six white horses carrying Parker to his grave in Kansas City; he makes Chan promise not to let his body be shipped back to K.C. for burial. What we're *not* told is that, against Chan's wishes, that's exactly what happened—to tell us this would require acknowledging that Parker was separated from Chan at the time of his death, and still legally wed to Doris Snydor, who is conveniently never mentioned in *Bird*. This movie is probably going to be praised in some quarters for its "unsensational" depiction of an interracial relationship. But the relationship between Whitaker and Venora could stand some sensationalizing. The only sparks that fly between them are acrimonious; they bicker from the word go. Although the script makes Chan an awful scold, Diane Venora brings unexpected shadings to the role: you believe in her as a thrill-seeking hipster who's just as glad when motherhood forces her into a more conventional way of life. Venora's is the film's only convincing performance. Michael Zelnicker is affectless as Red Rodney, the white trumpeter who sang the blues in order to pass as a black albino while on tour with Parker in the segregated South in the late forties. Zelnicker wouldn't have had to worry about white sheriffs; black audiences of the period would have hooted this yuppie off the stage. (He and Parker play a Jewish wedding, and when the cute little rabbi says about Parker and the other sidemen, "These boys are not Jewish, but they are good musicians," you feel as though you've witnessed this scene in a hundred other movies. Eastwood and Oliansky are delivering a sermon on the need for unity among oppressed minorities, and what's unbearable about it is that they think they're being subtle.) As the young Dizzy Gillespie, Sam Wright is sanctimonious and old before his time, and (as the jazz critic Bob Blumenthal has pointed out) the audience that knows nothing about the real-life Dizzy is going to wonder how he ever got that nickname. The first time we see Diane Salinger as the Baroness Nica, she's wearing her beret at a tilt that casts half of her face in shadow, and she watches Parker with predatory eyes. She's a shady lady from Grand Guignol. Why this visual insinuation about a woman who made her apartment into a salon for black musicians with whom she maintained platonic relationships, and whose only possible "crime" was that of dilettantism? She had no responsibility—symbolic or otherwise—for Parker's death.

The music in *Bird* has a phoney ring to it, even though Parker's recordings were used for most of the soundtrack. There were fans who used to follow Parker around the country, sneaking cumbersome wire recorders into nightclubs to preserve his work and shutting them off when his sidemen improvised. Eastwood and music supervisor Lennie Niehaus go these ornithologists one better (or one worse) by filtering out Parker's sidemen altogether in favor of new instrumental backing. In addition to being unfair to Parker's sidemen, many of whom were indeed capable of keeping pace with him, this removes him from his creative context and gives no sense that bebop was a movement. *

But Parker is out of context throughout *Bird*. The movie would have us believe that he had little curiosity about the world beyond jazz, which in turn showed only oppositional interest in him. In reality, the musicians who worshiped Parker remember him as well-read, with a consuming interest in twentieth-century classical composition. And black jazz musicians of Parker's era had a direct influence on those white artists from other disciplines—the nascent hipsters and beats who people such early fifties novels as Chandler Brossard's *Who Walk in Darkness* and John Clellon Holmes's *Go*. Parker was a source of fascination to these poets, novelists, and abstract impressionists who were beginning to define themselves as outlaws from middle-class convention. They recognized his artistic drive and suicidal self-indulgence as the *yin* and *yang* of a compulsive nature pushing against physical limitations and societal restraints. In *Bird*, few white characters, except those from the jazz underground, seem to know or care who Parker is, and he isn't sure himself.

The pity of all this is that Clint Eastwood is a jazz fan, and *Bird* is supposed to have been a labor of love. In 1982, Eastwood directed and starred in *Honkytonk Man*, the gentle, admirably straightforward story of a Depression-era Okie troubadour called Red Sovine, who succumbs to tuberculosis before realizing his dream of performing at the Grand Ole Opry. Among its other virtues, that film managed to suggest the succor that music can give both performers and audiences. Perhaps believing that Parker was subjected to a harsher reality than the fictional Sovine by virtue of being black and a drug addict, Eastwood's tried to find a more insistent rhythm for *Bird*, but the one he's come up with feels choppy, disconnected, and pointlessly arty, with dated experiments in time and point of view forcing him against his best natural instincts as a storytelling director.

Charlie Parker first appeared on screen in the guise of Eagle, a heroin-

---

* When I wrote this, I hadn't yet heard the *Bird* soundtrack. Read on.

addicted saxophonist played by Dick Gregory in the forgotten *Sweet Love Bitter* (1967), which was based on John A. Williams's novel *Night Song*. Although Gregory's performance was surprisingly effective, Eagle was a peripheral figure in a civil-rights-era melodrama about a white liberal college professor on the run from his conscience. In the late 1970s, Richard Pryor was supposed to star in a film about Parker that never got made—which is probably just as well, because Pryor brings so much of his own persona to the screen that Charlie Parker would have gotten lost. That leaves us with *Bird*, a jazz fan's movie in the worst possible sense—a movie with the blues, a *Birdland, Mon Amor* that wants to shout "Bird lives!" but winds up whispering "Jazz is dead." *Bird* communicates the melancholy that every jazz fan feels as a result of the music's banishment from mainstream culture. In projecting this melancholy on Charlie Parker—whose music still leaps out at you with its reckless abandon, and whose triumph should finally count for more than his tragedy—Eastwood has made another of those movies that make jazz fans despair that mainstream culture will ever do right by them or their musical heroes.

(November, 1988)

DECONTEXTUALIZIN' THE BIRD

Roger Ebert, who voted thumbs up, and Gene Siskel, who voted thumbs down, were on television debating the merits of *Bird*, when Siskel cut through the give-and-take to make an especially valid point: *Bird* tells us that Parker's music was great, but doesn't indicate the criteria by which greatness is measured in jazz. "What would you have wanted? A panel of jazz critics talking about Charlie Parker?" Ebert quipped. "When people see the movie, they'll be able to hear his music for themselves. If they buy the soundtrack, they'll be getting a Charlie Parker record. What's wrong with that?"

Plenty, as a matter of fact. The *Bird* soundtrack (Columbia SC-44269) is a Charlie Parker record only in the sense that the colorized *It's a Wonderful Life* is a Frank Capra movie. Because Parker's career preceded the technological advancements that today's record buyers and moviegoers insist on, Clint Eastwood and Lennie Niehaus must have felt that doctoring Parker's recordings was their only reasonable alternative. It might have been easier to hire an alto saxophonist to dub for Forest Whitaker, but this would have undermined the film's attempt to convince contemporary audiences of the initial revolutionary impact of a style of jazz that has now represented the status quo for almost four decades. Forget that none of the saxophonists now playing in Parker's style could have matched his instrumental dexterity, harmonic imagination, or tonal vibrancy. The insurmountable problem would have been finding a saxophonist capable of playing like Parker without giving the impression of recycling forty-year-old licks.

But a Parker surrogate would have been preferable to the mess that Eastwood and Niehaus have made in tampering with Parker's music in a misguided attempt to bring it back to life. All they succeed in doing is making Parker sound more ghostlike, as a result of the noticeable drop in recording quality whenever he begins an improvisation (a flaw that's admittedly more obvious on disc than in the theater, where the eye frequently distracts the ear). As the jazz critic Stanley Crouch pointed out in another context, the term *solo*, frequently used to describe a jazz improvisation with rhythm accompaniment, is misleading in implying that an improvisation can be isolated from the environment in which it's played, "[without understanding] that those notes were being fitted into a tempo and into the mobile context of an ensemble." In removing such Parker sidemen as Miles Davis and Max Roach from his performances, *Bird* erases history. However far ahead of all but his best sidemen Parker might have seemed at times, bebop was a full-fledged movement with followers playing as sizeable a role as the leaders, and the pressure that Parker's accompanists must have felt in keeping pace with him contributed to the palpable excitement of his recordings. Understandably, Parker and his "new" sidemen never achieve comparable rapport. How could they, with forty years separating their visits to the recording studio? There was no way this undertaking could have yielded satisfying results, but a seemingly arbitrary choice of musicians further sealed its doom. John Guerin, a West Coast studio regular whose drumming behind Parker is alternately perfunctory or overbearing, had no business replacing Max Roach. Unlike Guerin, the pianist Monty Alexander is nothing if not consistent—he's always sweaty, busy, and cliché-ridden, supplying far too emphatic a rhythmic accentuation to Parker's lines. Even when the accompaniment is provided by musicians on Parker's general wavelength (including the pianists Walter Davis, Jr., and Barry Harris), the results are hopelessly anachronistic, because these present-day boppers are playing in a now somewhat dated style that Parker was still in the process of defining. They have Dolby and digital mastering on their side, but it's Parker who communicates a sense of risk, despite the fuzz around his horn. In his liner notes for the soundtrack, Leonard Feather writes, in a classic bit of pretzel logic, that the album gives Parker a chance to play "alongside men whose company he never lived to enjoy." The pleasure was all theirs, as Feather—a personal friend of Parker's, and one of his earliest supporters—should be the first to concede.

The album's worst desecration occurs on "Ko Ko," and not merely as a side effect of Guerin's metronomic time-keeping. Parker and Roach were the only soloists on the original "Ko Ko," and it would have been difficult to imagine anyone following them. But on *Bird*, the track has been extended to include solos by Walter Davis, Jr., and the trumpeter Jon Faddis, with a predictable loss of intensity and compression. At least one can still find the unspoiled "Ko Ko" on various Muse reissues. Parker's original version of "April in Paris" with strings is likewise still available (without Niehaus's disingenuous bossa-nova "updating") on Verve. Not so the versions of "All of Me" and "This Time The Dream's on Me"—tapes from Chan Richardson's collection that document one of Parker's few recorded meetings with the pianist Lennie Tristano, recorded in Tristano's apartment with the drummer Kenny Clarke keeping time on wire brushes on a telephone book. If released as originally recorded, these informal tracks would be an invaluable addition to Parker's discography. But they're available only on the soundtrack, with Tristano and Clarke giving way to Alexander, Guerin, and Ray Brown, as in the film. In their present condition, these performances are disorienting, to say the least. Parker's melodic phrases are lengthier than usual, clearly in response to Tristano, but we don't hear Tristano—just Alexander, whose cluttered phrasing is the antithesis of Tristano's. Here's hoping for a bootleg.

In the movie, Parker is shocked to find that the clubs that once lined Fifty-second Street between Fifth and Sixth avenues in New York have been replaced by strip palaces and rock 'n' roll joints. This is supposed to be 1955, but the world has changed so much since then that it's possible to be nostalgic about strip palaces and rock 'n' roll joints. That block of West Fifty-second Street is now dominated by Warner Communications (which is distributing the film) and CBS (still the headquarters for Columbia Records, although the label is now owned by Sony). If Parker were to find himself on Fifty-second Street now, would he shudder to hear what these communications giants have done to his music? I'd like to think so, but I hesitate to indulge in such conjecture, because that's what led the makers of *Bird* to perpetrate this atrocity. "If [Parker] were among us today, this is unquestionably the way he would want to sound," writes Feather. If Charlie Parker, who was born in 1920 and died in 1955, were among us today, he would be sixty-eight. To pretend to know more than that is arrogant folly.

(November, 1988)

OBSESSION

Bruce Weber's *Let's Get Lost* is an extraordinary two-hour, black-and-white film portrait of Chet Baker, the trumpeter who fell to his death from a third-floor window of an Amsterdam hotel May 13, 1988, shortly after Weber finished shooting. Only the way that Baker died was surprising (the Dutch police ruled his death a suicide, but the U.S. Consulate declared it an accident, and Baker's intimates believe that he was murdered). Baker, who was fifty-eight, had been a drug addict for most of his adult life, and he looked like he was dying from a cumulative overdose the last time I saw him perform, at Fat Tuesday's in New York in 1986. He also looked like he didn't particularly give a shit. Once the most photogenic of jazz musicians, he still had his hair, and was still attractive in that lined and ravaged way that's frequently described as "rugged"—though not in his case. The lower half of his face was caving in (looking at him, I thought for a moment that he'd forgotten his dentures—he'd lost his teeth eighteen years earlier, in a drug-related mugging); all that was left of him were his wary, cooly appraising eyes. He spoke slowly and without inflection, as though in a permanent narcotic stupor, exactly the way he does in *Let's Get Lost*. "What's your favorite kind of high?" Weber asks him at one point. "The kind of high that scares other people to death," Baker answers without hesitation, his face a Nosferatu mask. "I guess they call it a speedball. It's a mixture of heroin and cocaine. . . ."

The obits explained Baker's less-than-brassy trumpet style in terms of a specific era and locale, crediting him with helping to spawn "cool" as a member of Gerry Mulligan's pianoless quartet in Los Angeles in 1952. But Baker's greatness was a matter of essence, not consequence: his ability to inhabit a melody was more important than his role in any movement or school. People will tell you that he played with *heart*, and as vague and romantic as the word sounds, it's the right word for what Baker projected: the white jazzman's equivalent of soul. His knowledge of chord changes was rudimentary, and his projection was weak. But with no flash to hide behind, he had to make every note count, and by the end of his life, his solos had become as deep and indelible as the lines of his face.

The current party line among jazz critics, in the wake of Albert Murray's influential *Stompin' the Blues*, is that jazz, like the blues that spawned it, is a celebration of shared cultural values between performers and audiences. Anyone who's heard Louis Armstrong or Charlie Parker knows that this is generally true. But anyone who's heard Bix Beiderbecke, Pee Wee Russell, Lester Young, or Paul Desmond knows that jazz can also be a vehicle for the expression of isolation and melancholy. It may be significant that all these soloists, with the exception of Young, were white men playing a black-identified music. Theirs was a song of isolation rather than community, and this is the lineage to which Baker belonged.

Music was only part of his appeal in the mid-fifties, when he was finishing ahead of Louis Armstrong, Miles Davis, and Dizzy Gillespie in the jazz-magazine polls. He was often said to resemble Montgomery Clift, James Dean, or Marlon Brando; the Hollywood screenwriter Lawrence Trimble—the most articulate of the "witnesses" in *Let's Get Lost*—likens Baker to PFC Robert E. Lee Prewitt, the moody army bugler played by Clift in *From Here to Eternity*. Baker's resemblance to these actors (and, later, to Clint Eastwood) was more than a matter of laconic delivery and a good jawline; like them, he was, in Trimble's apt words, "a slightly antisocial role model to look up to," in an era of college football heroes. What audiences responded to when listening to the young Baker or watching those actors, paradoxically, was the private nature of their blues.

Baker also sang, and, until nicotine dyed his timbre a manly brown, the adjective most frequently used to decry his crooning was "epicene." No one thought he was gay, but he sounded effeminate to some—an equally grave offense in the 1950s, a testosterone-counting decade in which the pianist Horace Silver's denigration of West Coast cool as "fag" jazz was widely and approvingly quoted, and in which two men meeting for the first time each felt obliged, as Norman Mailer once put it, to prove he was "less queer" than the other. But it was Baker who was in touch with the sexual tenor of the times: androgyny is now recognized as having been central to the appeal of Brando, Clift, and Dean, and most movie audiences probably recognized as much even then, if only subliminally.

If this talk of sexual ambiguity seems tangential to *Let's Get Lost*, consider that Weber, a commercial photographer known for the superrealist eroticism of his ads for Calvin Klein and Ralph Lauren, used Baker's early vocals as mood music for most of *Broken Noses* (1987), a wet dream of a documentary about a self-adoring Adonis of a prizefighter named Andy Minsker and the school-age Adonises-to-be he's grooming in the manly arts (but mostly just grooming, from what Weber shows us). In a written prologue to *Broken Noses*, Weber tells us that Minsker is a dead ringer for the young Chet Baker, to whom the film is dedicated. Though both of Weber's films are about unequivocally hetero men (and though his own sexual preference is, of course, none of our business), both films are homoerotically charged, and both demonstrate the extent to which the gay male sensibility that pervades the fashion professions (including commercial photography) has eroticized what even straight men see when they gaze at other men.

Weber tries your patience in *Let's Get Lost* (named after an escapist pop tune that Baker recorded in 1955, though the title also suggests perdition). Minsker, the punchy hunk from *Broken Noses*, is here for no apparent reason, cavorting on a Santa Monica beach with others from Weber's retinue, including a bonbon named Lisa Marie (described in the press kit as "a voluptuous young actress/model") and a jazzbo poseur called Flea ("member of the rock group The Red Hot Chili Peppers [and] a Chet Baker lookalike and fan"—and if you buy that, I got a vice-president for you who's the spittin' image of Robert Redford). Minsker and Flea, who condescend to Baker as an admirable old stud, are a reminder that we're on Weber's turf, not Baker's. So is the poster of Jean-Paul Belmondo in *Breathless* behind Baker during the interview sequences in Cannes, where Weber is presumably showing *Broken Noses*. There's a poignant moment at Cannes when Baker has to plead for silence from a crowd of second-string glitterati before singing and playing Elvis Costello's "Almost Blue." But instead of being moved, you're just angry: why did Weber drag Baker to Cannes in the first place? Shouldn't Weber be bird-dogging his subject, instead of the other way around? (If Weber remade *Nanook of the North*, Nanook would be wearing Calvins). Weber's self-reflexiveness almost spoils what should have been a powerful final scene in which Weber (off-camera) tells Baker how "painful it's been to see you like this" (strung out on heroin for five days, before methadone could be obtained for him). "Will you look back on this film in years to come and remember it as good times?" Weber asks Baker, who—understandably flattered by the attention he's been receiving as the subject of a million-dollar documentary, but otherwise opaque—asks how the hell else he could be expected to feel.

But even when you're staring as *Let's Get Lost* in disbelief, you can't take your eyes off it—a sympathetic reflex, in part, because you know that Weber can't take his eyes off his beautiful loser of a subject. It's an autobiographical film about somebody else (every journalist who's ever become hopelessly wrapped up in one of his subjects will know what I mean). "I remember, for the first time, [knowing] what photogenic meant, what star quality meant, or charisma," William Claxton, whose early album-jacket photos of Baker helped to make him a star, tells Weber—and you know that Weber experienced palpitations in first looking at those photos (like the narrator of Mishima's *Confessions of a Mask* looking at drawings of St. Sebastian). Jazz fans are going to complain that *Let's Get Lost* is more about Baker's mystique than his music, but the mystique was what drew Weber to the music, and the movie is finally as much about Weber's Chet Baker fixation as it is about Chet Baker—it's that fixation that gives this meditation on the nature of cool (replete with 1950s-style lower-case graphics) its sexual heat. Without the intensity of Weber's gaze, *Let's Get Lost* would be dull hagiography, like many of the documentaries shown at jazz film festivals and on PBS. When it's over, you know you've seen a movie. Trimble could be talking for Weber when he says, "Jazz musicians had names like Buck and Lockjaw and Peanuts and Dizzy, and he was just named Chet, which was sort of a soft sound. . . . The way he played, what he looked like, his name, everything—it all went together."

So could the trumpeter Jack Sheldon, who complains, good-naturedly, that "[Baker] didn't know what note he was hitting. He just pressed the valves down. It was *easy* for him. . . . I played the trumpet, too, and it was real *hard* for me." You get the feeling that Baker's unnegotiable independence is a large part of his attraction for Weber, who comes from a world in which making the right contacts and being seen in the right places is paramount. (And in contrast to Minsker and Flea, whom Weber no doubt spent hours dolling up to resemble Baker, the genuine article probably spent as little time worrying about how he looked as he did practicing.) When Weber badgers Baker's mother to admit that Baker disappointed her as a son (one of two instances when he ignores an interview subject's request to go off the record), the scene is moving because you know that Baker ultimately disappointed Weber, too: though he lived fast, he didn't die young and leave a good-looking corpse. The Chet Baker we see under the closing credits is the dreamboat from *Ulatori alla Sharrar*, a 1959 Italian quickie so obscure it's not listed in David Meeker's definitive *Jazz in the Movies*. (Nor is *Hell's Horizon*, another delightful curio Weber has dug up—a 1955 B-movie about the Korean War, with a surprisingly callow and athletic Baker as the kid in a battalion that also includes the Beaver's dad and Jerry the dentist from *The Dick Van Dyke Show*.)

Weber also interviews Baker's estranged third wife and her three grown children by him in Stillwater, Oklahoma, and two of his lovers: the singer Ruth Young, a tough cookie who disputes Baker's version of the beating that cost him his teeth, and characterizes his wives as "the crazy, the frigid, and the Virgin Mary" (she herself is "the bitch"); and Diane Vavra, a pathetic burn-out who tells how he beat her and once stood her up, and who flatters herself that he needed her to help him figure out how to phrase lyrics in the recording studio. This is riveting stuff, but *Let's Get Lost* is most seductive when no one on camera is speaking: when light and shadows and cigarette smoke are playing across Baker's aged face, or when, to the accompaniment of Baker's youthful singing, Weber and the cinematographer Jeff Preiss pan old album covers, publicity stills, and baby pictures. These deliriously shot sequences, which are tactile in their intimacy, flaunt something discreetly hinted at in Francis's worship of Dale Turner to the exclusion of all other jazz musicians in *'Round Midnight*: fandom as another name for fetish.

Chet Baker walks through *Let's Get Lost* in a daze, though he seems to be enjoying the antics of Andy Minsker and Flea. The scenes in which Baker seems most alive are those in which he's shown recording the movie's soundtrack album—especially when he's whispering lyrics close to the mike, shaping the words with his mouth. Or at least, that's the way it looks in the movie. But the soundtrack (RCA Novus 3054-1-N) is the sort of album you're reluctant to listen to very closely—not because it doesn't reward undivided attention, but because you're afraid it'll get to you. It isn't just that the tempos are so funereal. Baker's trumpet tone is slack, his vocal phrasing is labored and gasping, and his sense of time is uncertain (though the pianist Frank Strazzeri gallantly tries to keep him in proximity to the beat, and just as gallantly covers up for him when he falters.) Wobbling between the diatonic, the pentatonic and the catatonic, Baker has never sounded worse—and no one has ever sung better. He was an intuitive musician, and these battered interpretations of "Every Time We Say Goodbye," "You're My Thrill," and "Imagination" show that intuition, unlike technique, is something that can't be lost. The album's final track is Baker's only recorded version of "Almost Blue," which Elvis Costello says he wrote while thinking about Baker. In the movie, Baker pleads with the revellers at Cannes for quiet, explaining that "It's one of those songs." It certainly is. It's so enveloping, in its almost inaudible way, that nothing could possibly follow it. You don't even feel like putting another record on.

(April, 1989)

AVANT-GARDE COMRADES

"There was always something a little off, like a Russian playing jazz," muses Fielding Pierce, the narrator of Scott Spencer's latest novel, *Waking the Dead*. Although the character is describing an uncharacteristic show of solicitude toward a disenchanted lover, the author is appealing to a territorial prejudice harbored by most Americans, including millions who know nothing about jazz save that it's indigenous to the United States. In fact, Russians were playing ragtime by 1910, if not before. Moreover, they have persisted in playing every jazz offshoot, from dixie to free, despite periodic opposition from the Kremlin, which has yet to decide, once and for all, whether to endorse jazz as an oppressed race's cry for freedom or to condemn it as a decadent Western consumer good. (Marxist ideologues have rarely known what to make of popular culture.) For decades it was assumed in the West, with no generally available recordings to the contrary, that Russian jazz was crude and inferior— a little off, in other words—because Russian musicians had had no direct exposure to the music's black progenitors. But S. Frederick Starr, in his *Red and Hot: The Fate of Jazz in the Soviet Union, 1917–1980*, published in 1983 by Oxford University Press, turned up compelling new information subsequently corroborated on Harlequin Records' *Jazz and Hot Dance in Russia, 1910–1963*. Jazz scholars should now concede that by the late 1930s, there were at least three Soviet musicians equal to any elsewhere in Europe (except Django Reinhardt), if still far behind those in the United States: the bandleader Leonid Utjesov, the pianist Alexander N. "Bob" Tsfasman, and the German-born trumpeter Eddie Rosner. Even before the publication of *Red and Hot*, it was common knowledge that certain Soviets were proficient in bebop: the trumpeter Valery Ponomarev, for example, who joined Art Blakey and the Jazz Messengers after emigrating in the mid-1970s. (Though still active, Ponomarev has become the answer to a trivia question: whom did Wynton Marsalis replace on trumpet with the Messengers?) Despite impressing American musicians and critics with their craftsmanship, these Soviet boppers scored no points for originality; although, to be fair about it, bebop for some time now hasn't been an idiom from which one expects much originality. Fortunately, there's more to contemporary jazz than bebop, and because the jazz avant-garde borrowed so heavily from alien sources in the aftermath of Ornette Coleman and Cecil Taylor, it became reasonable to assume that the Soviets might finally have something of their own to contribute to what was rapidly becoming a global art. Still, it took many in jazz by surprise when the Soviets, as part of the cultural accords struck by Reagan and Gorbachev at Geneva, agreed to export an avant-garde jazz group along with such familiar commodities as the Leningrad Symphony and the Bolshoi and Kirov ballets.

The Ganelin Trio, from Lithuania, last summer became the first Soviet jazz group to tour America. Their three-week tour started in New York on June 21, and took them to fourteen other cities, including Vancouver and Toronto. This wasn't the first trip West for the group, which is made up of the pianist and composer Vyacheslav Ganelin, the saxophonist Vladimir Chekasin, and the percussionist Vladimir Tarasov. They'd toured Great Britain in 1984, performed at jazz festivals in West Germany and Holland, and been sent to help solidify Soviet relationships with Communist parties in France, Italy, Austria, and Portugal. Until last summer, though, only a handful of Americans knew the Ganelin Trio's work, and only through records. Two albums that the group had made for Melodiya, the state-owned label and the only label in the U.S.S.R. for which Soviet musicians are officially permitted to record, had reached the West through leasing arrangements with jazz specialty labels. But the albums that had aroused the most curiosity about the Ganelin Trio were nine released without Soviet consent by Leo Records, a British shoestring label operated by Leo Feigin, a Soviet émigré. Feigin says that he's not at liberty to disclose how he came into possession of the Ganelin Trio's concert tapes, except to absolve the musicians themselves of complicity.

I heard the Ganelin Trio in Philadelphia and San Francisco, and in both cities they turned in spectacular performances that confirmed the impression the Leo albums had given of them as one of the world's premier avant-garde jazz ensembles. Although I (and everybody else) have long heard that Soviet artists are basically no different from their Western counterparts, I was still struck by how much, in conversation as well as in performance, Ganelin, Chekasin, and Tarasov, all of whom are around forty, resembled their contemporaries in the United States and Western Europe. Before the concert in San Francisco, I asked Tarasov—the trio's only English-speaking member—to name his favorite contemporary classical composers. "Other classical composers?" he asked, pointing to Ganelin before listing Cage, Berio, Stockhausen, Penderecki, and Alfred Schnittke. Substitute the names Glass and Reich for Schnittke, a Soviet composer of little influence in the West, and Tarasov's answer is one you might expect from a sideman with, for example, Anthony Braxton or Anthony Davis, to cite just two U.S. musicians categorized as jazz composers, although they hold dual citizenship in jazz and avant-garde classical music. Significantly, Ganelin is the musical director for the Russian dramatic Theater in Vilnius, Lithuania; Chekasin teaches music at the Vilnius Conservatory; and Tarasov, though self-taught, works full-time with the Lithuanian State Symphony. These are jobs of the sort that vanguard musicians in the west either have or covet. Ganelin has described the trio's music as "polystylistic": a way of saying that he juxtaposes not only jazz from various eras, but classical and ethnic musics as well, in harmony with the recent eclectic tendencies of Western avant-garde composers. Ganelin, Chekasin, and Tarasov are said to have declined an invitation to jam with American musicians at a reception at Gracie Mansion before their New York concert, apparently fearful of being misjudged for their lack of intimacy with bebop and the blues. I know many younger American musicians who might also have declined such an invitation, for the same reason.

Although the Ganelin Trio's music has a boisterous, at times even bellicose, tone, in all the sets I heard them play, the emphasis was on disciplined motivic exposition rather than spontaneous expression, as in the best recent American jazz. Each set consisted of one opus, as a set by Cecil Taylor might. As a pianist, Ganelin resembled Taylor in stamina, speed, and intensity, though his rhapsodic lyrical passages were more akin to the musings of Keith Jarrett or the Romantic flourishes of Ganelin's countrymen Sergei Rachmaninoff and Vladimir Horowitz. Ganelin's percussive density freed Tarasov to concentrate on tintinnabulation rather than propulsion, and the drummer tapped out bewitching rhythms on every object he laid his hands on—bells, shakers, chimes, even a tin water cup, in addition to his traps. From time to time, Ganelin and Chekasin also picked up small percussion instruments and a variety of pots and pans, in a manner reminiscent of the Art Ensemble of Chicago and other black groups affiliated with the AACM, albeit without the African signification.

Like the members of most of these AACM groups, Ganelin, and Tarasov are multi-instrumentalists. In fact, they often play two or more instruments simultaneously; Ganelin producing chimerical bass lines and atmospheric special effects from a variety of small electronic keyboards stacked atop the piano, à la Sun Ra, and Chekasin—who plays with a crazed vehemence suggestive of the late Albert Ayler and the German avant-garde Peter Brötzmann—creating skewered overtones by blowing through two saxophones at once, à la Rahsaan Roland Kirk. There's an air of post-Beckett theatricality to the Ganelin Trio's multi-instrumentalism, a suggestion of mime in Chekasin's arm-waving and arrhythmic foot-stomping, and a touch of absurdist buffoonery in their send-up of "Mack the Knife," the tune they offered as an encore both times I heard them. In this respect, as in many others, they bear greater similarity to the Willem Breuker Kollektief and the Misha Mengelberg–Han Bennink Duo than to musically comparable black American groups, many of whom still associate broad humor with Uncle Tom.

Despite these many points of comparison to Western counterparts, there was nothing back-numbered or secondhand about the Ganelin Trio's brand of jazz. The source of the trio's originality is hard to place. In the absence of recognizable Soviet folk interpolations or statements of purpose from the musicians themselves, one has to accept on faith the contention of Leo Feigin and others that the trio's music owes its special fervor to its courageous, if tacit, rejection of Soviet socialist realism in favor of a long-suppressed Slavic penchant for free-spirited abstraction. This, according to Feigin, is what links the Ganelin Trio to Stravinsky, Mayakovsky, Diaghilev, Chagall, and Kandinsky.

The Ganelin Trio's North American tour received more coverage than any other event in jazz this summer, with the possible exception of Benny Goodman's death. There was attention even from *The Today Show* and *Entertainment Tonight*, television shows not usually so hospitable to avant-garde jazz. As a result of this publicity and the natural curiosity surrounding any Soviet export, the Ganelin Trio's concerts drew many ticket-buyers previously unexposed to avant-garde jazz, which probably explains why—in Philadelphia in particular, but even in San Francisco, where ROVA, an avant-garde saxophone quartet with its own local following, was also on the bill—there were audible sighs of relief followed by loud bursts of applause whenever the musicians happened into a steady four beats per measure.

Some American jazz critics suggested in their reviews of the concerts that the Ganelin Trio might pique interest in homegrown avant-gardists. This is an outcome much to be desired, but one that strikes me as improbable, for reasons best summed up in the reaction of a friend of mine, a journalist who thinks of herself as liking jazz though she rarely buys records or goes to concerts. "I'm glad I heard them, but I wouldn't want to hear them again, and I do wish they had played a little more American jazz," she told me—and I'd guess that her reaction to Cecil Taylor, Anthony Braxton, or the Art Ensemble of Chicago might have been the same. And inasmuch as novice listeners often mistake passion for anger in free jazz, many of those in the predominantly white crowds that showed up for the Ganelin Trio probably felt less threatened in the presence of comrades (perceived target: Soviet bureaucracy) than they would have in the presence of brothers (perceived target: whitey).

Less than a week after the Ganelin Trio returned to the Soviet Union, an episode of the BBC-produced *Comrades*, shown here on PBS last summer, was devoted to the tribulations of Sergei Kuryokhin, a self-absorbed Leningrad-based avant-garde jazz musician who's begun to toy with rock 'n' roll. Much was made of the fact that Kuryokhin was "unofficial"—that is, not a sanctioned member of the Composers' Union (unlike Ganelin), and therefore not permitted to record for Melodiya or to advertise his concerts, except by word of mouth. What wasn't mentioned, but perhaps should have been, if only in the panel discussion that followed, was that jazz visionaries are also "unofficial" in the United States, at least in the sense of being caught between popular culture and the fine arts, unequipped to compete for the consumer dollar yet allocated only a fraction of the institutional funding lavished on artists in other disciplines. Kuryokhin probably has a better chance of eventually recording for Melodiya than the typical American jazz avant-gardist has of recording for CBS or any of the other corporate-owned major labels. If anything, the Ganelin Trio's example proves that the Soviet jazz avant-garde has been more successful in attracting a mass following. It's impossible to imagine an album by David Murray or the World Saxophone Quartet selling 65,000 copies, as each of the Ganelin Trio's Melodiya albums has in the U.S.S.R. It's also impossible to imagine our State Department sending an avant-garde ensemble on a U.S.S.R. tour: to judge from past history, Washington's enthusiasm for jazz is limited to big bands and safely assimilated veterans. (ROVA did go to the Soviet Union in 1983, but at their own expense, without government sponsorship.)

Of course, it's better not to push the comparison between the musician's life in Leningrad and the musician's life on New York's Lower East Side further than it should go, because Kuryokhin's starving-artist equivalents in the U.S. can better afford the luxury of *la vie bohème*. In San Francisco, the Bay Area Council for Soviet Jews distributed leaflets outside the Ganelin Trio's concert, without attempting to dissuade patrons from entering the hall. John Ballard, a Wyoming concert promoter who brought the Ganelin Trio to North America, introduced himself to the protesters, bought one of their T-shirts demanding the release of imprisoned Soviet Jews, but understandably declined their offers of free shirts for the musicians. It's much safer to protest discrimination against Soviet Jews in San Francisco than it would be in the Soviet Union, especially for Vyacheslav Ganelin, who is himself Jewish. Even though the Ganelin Trio was traveling without a KGB chaperon, the members shied away from all questions with political implications, including one I tried to sneak past Tarasov about whether he thought the group's clandestine Leo releases accurately represented their music.*

In the U.S., musicians are free to inveigh against racism, the oligarchic record business, and anything else they perceive as oppressive. But who listens to them? More to the point, who listens to their music? The avant-garde has it worst of all, but sometimes it seems as though all jazz musicians, past and present, are *personae non gratae* in mainstream American culture. By coincidence, a week after The Ganelin Trio performed in Philadelphia, a local

---

* Leo Feigin said that when he arrived in New York for the Ganelin Trio's first U.S. performance, he was surprised to find the trio unchaperoned. As an voluntary exile responsible for releasing unauthorized Soviet jazz performances in the West, Feigin would have been officially off limits to the musicians if a KGB agent had been present, but he had come prepared to bribe the agent with an expensive cassette recorder. "I was absolutely sure there would be someone from the KGB, and I couldn't believe my eyes when I saw that there wasn't," Feigin said. "I guess that the Soviets are clever enough to realize that the Ganelin Trio has traveled so much already, if they wanted to defect, they could have done it a long time ago."

But Vyacheslav Ganelin, the son of a party official, did, in fact, emigrate to Israel in 1987.

repertory cinema screened *Jazzman*, a sweet 1984 Soviet film set in the 1920s, about a Russian conservatory dropout so taken with the music of Scott Joplin and Duke Ellington that he decides to form his own band. Gjon Mili's famous 1944 short, *Jammin' the Blues*, with its opening overhead shot of Lester Young's trademark porkpie hat, preceded the main feature. As Young's face came into view, a man sitting in front of me turned to his companion and said, "I thought you said these guys were supposed to be Russian." For all most of us know about them, and their present-day descendants, they might as well have been.

Harlequin Records' *Jazz and Hot Dance in Russia, 1910–1963* (HQ 2012) can be ordered from Daybreak Express Records, P.O. Box 250, Van Brunt Station, Brooklyn, New York 11215. Of the nine Leo releases by the Ganelin Trio, *Vide* (LR-117) and *New Wine* (LR-112)—both uninterrupted album-length suites, the latter a series of sly variations on the Broadway show tune "Too Close for Comfort"—are the most impressive. But *Strictly for Our Friends* (LR-120) might serve as a better introduction, because of its greater variety and shorter performances. The Leo catalog also includes intriguing albums by Kuryokhin (LR-107), the saxophonist Anatoly Vapirov (LR-110 and LR-121), the pianist Harry Tavitain (LR-124), the singer Valentina Ponomareva (LR-136), and the groups Arkhangelsk (LR-135) and Homo Liber (LR-114 and LR-129), among other eastern European contraband. Leo albums can be ordered from New Music Distribution Service, 500 Broadway, New York, N.Y. 10012, and North Country Distributors, Cadence Building, Redwood, N.Y. 136789.

These distributors are also good sources for the Ganelin trio's *Non Troppo* (hat ART 2027), a two-record set including one of the group's "official" Melodiya sessions. In 1985, East Wind Records (3325 Seventeenth Street, N.W., Washington, D.C. 20010) released *Poi Segue* (MC-20647), the other Ganelin Trio Melodiya album to find its way to the West, along with four other Melodiya albums by derivative Soviet bebop and fusion groups, of interest only as arcana. The best fund of background information on the current Soviet jazz ferment is *Russian Jazz New Identity*, a collection of interviews and panegyrics (many of them impenetrable), edited by Leo Feigin and published by Quartet Books earlier this year.

(November, 1986)

BAND OF OUTSIDERS

When I replace Letterman, there'll be no more jumping from ladders into layer cakes and instant-replaying the mess—the high-tech, PG-rated geek show ends, and the World's Most Dangerous Bar Mitzvah Band has to go. The band I'm considering as a replacement is the Microscopic Septet, a New York saxophone-quartet-plus-rhythm whose riffs do what riffs are supposed to do: set your pulse racing and lodge in your skull for days on end. When my opening monologue falls flat (my jokes are based on the chord changes of old Henny Youngman jokes), I'll banter with Phillip Johnston, the Micro's soprano saxophonist and *de facto* leader, who talks to audiences in quotation marks, like Paul Shaffer, but with a more lethal twist of irony ("Please don't change; let's savor being together in this moment," he deadpanned between numbers at Visiones in Greenwich Village two Sundays ago. "Oops, somebody changed.") Some night, I'll get Johnston to tell the story of how he and Joel Forrester, the group's pianist and other resident composer, first met in the Bowery in the early seventies, shortly after Johnston dropped out of NYU, after only a few weeks of school. ("I accepted the scholarship because it got me out of Queens. I met Joel when I was practicing a Thelonious Monk tune, and a guy I had never seen before came walking through my door, which wasn't locked, because those were the hippy days . . ."). After Johnston upstages me with his remark that "when we formed the band around 1981, we wanted to call it Claude Dunston and the Psychic Detectives, but that was too unwieldy, so we considered the Microencephalic Septet," I'll turn to Forrester and say, "You actually knew Monk, didn't you, Joel?"

"Right," Forrester will say. "This is how I met him. I was once fired from the West Boondock, a chicken and ribs joint, for playing 'repetitive' music, and   ."

"Don't you mean minimalist music, Joel?"

"No, because it wasn't simple. Repetitive music, with far-out harmonies from Scriabin and Charles Ives. Anyway, a guy at the bar once threw up while I was playing, so I was fired. Every six months or so after that, I'd go and audition for my old job back. Once, while I was there for my semiannual dose of humiliation, I recognized the Baroness at the bar. . . ."

That would be the wealthy jazz patron and Rothschild-by-marriage, the late

Baroness Nica de Koenigswarter, I'll ask for the benefit of viewers watching us in Squaresville.

"Yes. For my audition, I played 'Pannonica,' the tune Monk wrote for her, and she used her influence with the owner to get me rehired. Then she took me off in her Bentley to Weehawken, New Jersey, where Monk lay fully dressed as though lying in state, but still alive. My 'sessions' with him amounted to this: the piano was just outside his bedroom, and if he liked what he heard, he got up and opened the door. If he didn't, the door stayed closed. Another consequence of that day was that the Baroness met my wife, whom she accused of not wanting to have a baby by me. Soon afterward, I noticed, our son Max was born."

Even without having Forrester tell the story of how, just out of prison for draft evasion and still on parole, he played for silent movies in San Francisco under the pseudonym Dr. Real (also the name of a character from one of his unpublished novels), I'm going to be upstaged a lot on *Late Night*, especially when I become so caught up in the Microscopic Septet's music that I forget to bring on my guests (which might be just as well, unless somebody does something about the Del Rubino Triplets and Brother Theodore). I think what delights me most about the Micros (the other members are the alto saxophonist Don Davis, the tenor saxophonist Paul Shapiro, the baritone saxophonist Dave Sewelson, the bassist David Hofstra, and the drummer Richard Dworkin) is their resistance to the easy temptation of eclecticism as an end in itself. This is quite an accomplishment for a band that resolutely packs so many non sequiturs into every single piece of music it performs. Take, for example, Johnston's "Waltz Of the Recently Punished Catholic School Boys"—from last year's *Beauty Based on Science (The Visit)* (Stash ST-276), the Septet's fourth and best album so far. It doesn't matter that the piece begins with Shapiro's delineating a melody so Hebraic in intonation it recalls Al Jolson (hey, at least it really is a waltz),* and ends with a sprightly tag that sounds like "Basin Street Blues" sped up into bebop. Nor does it matter that both the *Get Smart* signature theme and a pedal point resembling one from Charles Mingus's "The Black Saint and the Sinner Lady" figure prominently on "The Dream Detective," one of Johnston's two halfway-between-dreamy-and-bleary *noir* evocations on

---

*"Why is it called that?" Johnston repeated my question. "It tries to express the mournfulness deep in the progressive soul of a Catholic elementary school boy like I was, in short pants, with a clip-on tie. Catholic school is a very militaristic atmosphere. It discourages progressive thinking and creativity, but floating within you is a soul that wants to leap out. I had this vision of hundreds of Catholic school boys floating above the city, holding candles and. . . ."

"You see what happens in this band," Forrester interrupted. "Someone writes a simple, straightforward waltz about Catholic-school boys floating above the city, and when guys like Shapiro and Hofstra get their hands on it, it turns into something utterly other."

the album. (The other is "Rocky's Heart," which boasts an unfaltering Hofstra bass line, a powerfully moody Shapiro tenor solo, and a surprisingly light-hearted Latin bridge that arrives totally out of the blue.) What matters, finally, is that none of the album's tracks is a hopeless cobble, because the level of musicianship (best exemplified by the panting tempo toward the end of Forrester's "The Visit," and Dworkin's efficiency at effecting quick transitions) is so high that nothing sounds condescendingly tossed off.

The Micros rehearse together at least once a week regardless of whether they have a gig coming up, and this dedication to music and fellowship was obvious from the panache with which they interpreted Johnston's and Forrester's difficult scores at Visiones. Counting the arrangements of Monk tunes and the contributions of Ohio saxophonist (and honorary Micro) Bob Montalto, the band's book includes one hundred and thirty-five compositions, fewer than a quarter of which have been recorded. Of the unrecorded material played at Visiones, two Forrester charts were especially impressive: "Winter Thunder," which pivoted on fast-moving semiminimalist ("repetitive," Joel?) harmonies and Dworkin's galloping drums, and was as rich and evocative as good movie music; and "Money, Money, Money," which had a late-emerging sad-party theme, a pecking R & B alto solo by Davis (who had become a father two days earlier and was still passing out cigars), high-register squeaks and sub-tone blasts by the impeccable Sewelson (playing on a borrowed horn, his own having been stolen), and a raucous "Can't Turn You Loose" ending. Forrester, who was burdened with a small Roland electric piano but made it ring like an acoustic grand, spent a good part of this last piece on his feet, doing a hunched little dance step and goosing the band with his body English, like Monk used to do (and Sun Ra still does). In Johnston's pieces, Monk is refracted through Steve Lacy, who is also Johnston's chief influence as a soloist; I haven't heard another soprano saxophonist who comes closer to capturing the quizzical tilt of Lacy's lines. At one point, the rest of the band briefly dropped out while he and Dworkin duetted, and the terrific drummer matched him microtone for microtone on his cymbals. (It's impossible to overpraise Dworkin, who punctu-ates ensemble passages with the flair of a big-band drummer and the precision of a symphonic percussionist.)

I'd guess that many of this band's allusions are unconscious, the inevitable result of its members' having played so many different kinds of jazz and pop. (Hofstra, who also plays tuba on the Micros' more Ringling Brothers pieces, was the bassist on the Waitresses' "I Know What Boys Like"; Dworkin has drummed for both Bill T. Jones and the big-breasted West Coast stripper Carol Doda; Johnston gigs with SoHo art-noise and polka bands; and both he and Forrester admit to having played more than their share of tangos, merengues, and schlock medleys for celebrants in the outlying boroughs.) In its proud New York pluralism, the Micro's music bears a philosophical resemblance to John Zorn's who—not surprisingly—is an ex-Micro. But the Micro's acoustic orientation, blurrier juxtapositions, and commitment to jazz as a *language* rather than a vocabulary, result in a kinder, gentler postmodernism than Zorn's dystopian cool jazz *cum* speed metal.

So why aren't these guys rich and famous, or at least unanimously applauded by those in the know? As locals without national reputations, the Micros lack the mystique it takes to impress New Yorkers, and it's impossible to develop that mystique when you're down on your knees doing your own sound a half hour before showtime in a dump like Visiones. And I suspect that what alienates some are the very qualities I find so endearing in the Micros: their lack of sanctimony and the way their concise, stoptime horn solos function like breaks in early jazz, relating thematically not only to a tune's "head" but also to the many subheads to follow (even Sewelson's *Get Smart* quote, which I assumed, on first hearing it on record, was a spontaneous interpolation, turns out to be a compositional signpost). Concision isn't much valued by the sort of jazz fan who dotes on lengthy solos, but the Micros' white postgraduate humor might be an even bigger stumbling block in an era in which jazz is regarded by young black musicians and their audiences as deadly serious business. Some critics who should know better have even equated the Micros with the Lounge Lizards and bohemian "fake" jazz (it must have something to do with Dworkin's *Stranger Than Paradise* fedora and slack Richard-Widmark-as-Tommy-Udo facial expression). With their in-joke titles and punning liner notes by literary absurdists Richard Foreman, William Kotzwinkle, and Ron Padgett, the Micros are sometimes too clever for their own good. But on the bandstand, their humor is difficult to resist. This is a band that knows how to have fun while going deep, and one would think that, with proper exposure, that combination would give them widespread appeal. Somebody oughta put these guys on TV.

(*March, 1989*)

BRONX CHEER

Remember on TV game shows in the 1950s the gonzo cry from the audience whenever a contestant said he was from Brooklyn? This always mystified me when I was growing up in Philadelphia, and a friend who was originally from Sheepshead Bay says it mystified her, too. It probably had something to do with Ralph Kramden or the Dodgers. Thanks to BAM, Spike Lee, and a reputation among artists as a borough where you don't need a Guggenheim to pay the rent (at least not yet), Brooklyn now has a classier, avant-gardsy kind of image. But the shouting has started again; only this time it's my fellow jazz critics who are making rude, approving noises. And I still don't get it.

Brooklyn's large black population, its relatively manageable rents, and its proximity to Manhattan have long combined to make it a hospitable roosting place for jazz musicians. But if there was a trademark Brooklyn "sound," no one had a clue what it was, until Peter Watrous broke the news in the lead article of a *Village Voice* jazz supplement in August, 1987. Watrous (who I should admit at the outset is a friend, though maybe not after he reads this) cheered the eclecticism of a close-knit group of black Brooklyn-based musicians in their late twenties or early thirties who, though beginning to make names for themselves playing deep jazz, hadn't renounced the James Brown and George Clinton records they'd grown up dancing to. These musicians—including the pianist Geri Allen, the singer Cassandra Wilson, the trumpeter Graham Haynes, the trombonist Robin Eubanks, the alto saxophonist Greg Osby, the guitarist Jean-Paul Bourelly, and the drummers Terri Lyne Carrington and Marvin "Smitty" Smith—were, in fact, aiming for a synthesis of jazz, hip hop, world beat, and quiet storm that "if [they're successful], it's history, and everybody, everybody young that is, will have to learn [their] language or risk becoming an anachronism." According to Watrous, and to the musicians themselves, the movement's shogun was Steve Coleman, "Kung Fu freak, James Brown student and hip hop-o-phile, second-generation Charlie Parker groupie, structuralist and rule maker supreme, computer boss, and killer alto saxophonist . . . a theoretician who's deep into a fresh, fuck-offish style of improvisation and composition, proving to younger members of the group that there's a way out of bebop death."

Don't wait for me to explain what a "fuck-offish" style of improvisation and composition is. I'm just quoting. About a month later—right on schedule as these things go—the *Times* reported the same news in more temperate language. "These musicians brainstorm together, ignore pigeonholes, and upset apple carts," Jon Pareles wrote in a Sunday "Arts and Leisure" piece. "The Brooklynites cruise across categories with grace and sly humor." Raves from the *Voice* and *Times* don't necessarily sell records, but these publications are the most influential publications in the world in solidifying critical opinion. Toward the end of the year, a promotion man for a West German label specializing in ECM-like dreamscapes complained to me that he was having a difficult time placing stories close to home, because German critics were talking only about the Brooklyn school.

Needless to say, Coleman's *sine die* (Pangaea PAN-7709), his first U.S. release after three for the West German label JMT, received a euphoric greeting when it was released this June. (As one of the first releases on a label run by the rock star Sting, the album also had the stardust factor going for it. Let's recall that Branford Marsalis got more publicity for being Sting's sideman than he ever had for being Wynton's brother.) In playing devil's advocate, I'm not going to argue that *sine die* is meretricious product. On first listen, I liked the reedy ululations of Coleman and fellow saxophonist Gary Thomas, the infrequent brass blatts of Robin Eubanks and Graham Haynes, the stutter-step rhythms, the steel-drum and thumb-piano effects, and the deliberate clash of acoustic and electric instruments on some tracks. This isn't fusion: the bass lines are sharper-edged and more complex, and there's more formal sophistication in the ensemble writing, and less gratuitous technical display during the brief solos. The mix is crystal clear, with each instrument well defined—no small consideration in the recording of music so dependent on instrumental crosstalk. On that level alone, *sine die* is a quantam leap over *On the Edge of Tomorrow* and *World Expansion* (*By the M-Base Neophyte*), the two JMT albums featuring Five Elements, Coleman's eight-member jazz-funk band, which also performs here, along with such added starters as Branford Marsalis and former group-member Geri Allen.

So what's the problem? For starters, the vocals of Cassandra Wilson, the Brooklyn school's grand diva. Her homiletic lyrics put to a severe test the dictum (Pound's?) that what's too silly to be said can be sung. And as for her singing, my failure to hear the "sensuality" that certain of my colleagues have described in tumescent prose makes me wonder if I'm reaching the age where I'll have to start wearing the bottoms of my trousers rolled. * (Graham Haynes's

---

* Wilson's *Blue Skies* (JMT 834-419-2), released a few weeks after this was written, was the album that was supposed to change my mind about her. It didn't. It's an album of standards for jazz fans who don't like standards—you can tell that the singer doesn't like them, either.

lyrics to "Darkness to Light," though nicely sung by Wilson, are even worse.) On "Profile Man," Coleman delivers his own rap, and though it at least has the advantage of being more direct than the rest of the album's vague poetics, the combination of his Hendrix-like mumble and a beat–box rhythm track amounts to a mixed metaphor. If Wilson's vocals were *sine die*'s only shortcoming, they could be charitably overlooked. But the overriding problem is that a second or third listen reveals this music to be Jazz and Funk Together Again for the Very First Time, with keyboard vamps recycled from Miles Davis (circa *On the Corner*) and short, torrid saxophone licks like those Ornette Coleman (no relation) has been playing for more than a decade with Prime Time (the difference, of course, is that when Ornette plays them, they aren't just licks). From what I read, Steve Coleman's study of the martial arts and other aspects of Japanese culture is supposed to have a profound bearing on his music, but I don't hear this, except in a negative sense: given *sine die*'s urban din, Coleman's solos want the bubble lettering of graffiti, but deliver the thin brush-stroke of Japanese calligraphy. Originality aside, the best reason for playing any style of music is that you can. But I'm afraid that Coleman, whose chief assets elsewhere (most notably on his records as a sideman with the bassist Dave Holland) have been a rarified tone and a slippery, approach-avoidance relationship to the beat, just isn't a very convincing soul man.

The fuss about Brooklyn in general and Coleman in particular reflects a crisis in jazz criticism. In his review of *sine die* for *Spin*, Gene Santoro concluded that this album and others in a similar vein demonstrate "that jazz . . . may be useful now primarily as a historical description. Louis Armstrong and Coleman Hawkins played jazz, for instance; so did Bird and Diz. But how do you categorize the music of Miles Davis from the mid-sixties on? Or of Ornette Coleman, especially with Prime Time?" I make a point of not answering rhetorical questions, but Santoro's strikes me as especially irrelevant, because I know some moldy figs who don't consider even Hawkins jazz, much less Parker and Gillespie. The question of what is or isn't jazz is as old as jazz itself (older, when you consider that the jury's still out on ragtime). The curious new twist, however, is that critics are now likely to withhold the designation from music they endorse, not music they deplore. This might be because so many younger jazz critics, whether out of economic necessity or honest eclecticism, spend as much time, if not more, writing about pop and world beat, and jazz just doesn't seem exotic or sexy enough to them. In part, this would explain why they've rallied behind Steve Coleman and his M-Base-hip-hop-samurai culture club, who evidently feel the same way.

But there are other factors at work, including a lack of hard-won skepticism in the young, and (as the close association between Wynton Marsalis and

Stanley Crouch illustrates) the prestige awaiting the first critic to spot a rising star (it may have been coincidence, but Watrous graduated to the *Times* soon after his Brooklyn piece). Says a Manhattan-based pianist sometimes associated with Brooklyn as a result of playing with Coleman and Greg Osby: "I like the publicity, but I have mixed feelings about it. I'm a little older than most of them, and I'm more tied to the acoustic tradition. What I like about what's going on in Brooklyn is that these younger musicians are exploring together and writing for one another's bands. There should be more of that in New York. But it's no accident that certain people are getting all that attention. Some of them are very savvy that if they present themselves to the media as a contingent, the media will be more interested in that than they would be in them as individuals. But believe me, as individuals they have very different approaches." As this comment suggests, the most insidious factor of all may be pressure from editors to make every piece on jazz, even if it's only a record review or profile, pinpoint a coming trend. The *Times* is guilty of this in all of its arts coverage, but especially in jazz and pop. Why can't editors who insist that critics tell readers where jazz is heading accept the fact that jazz is heading everywhere at once and nowhere in particular, just like every other art at the end of this century to end all centuries? Wasn't it just five years ago that everybody was touting New Orleans and the swing back to tradition, as epitomized by Wynton Marsalis? I'm reminded of a recent Roz Chast cartoon for *The New Yorker*. It showed a bunch of dazed travellers, baggage in hand, standing beneath a board that read "Next Bandwagon Leaving. . . ." I don't remember the time of departure, but it's unimportant. The point is there's one leaving every hour.

(October, 1988)

BORN OUT OF TIME

Just eighteen when he made his debut with Art Blakey and the Jazz Messengers, the trumpeter Wynton Marsalis answered the prayers of those who feared that the clock was running out on jazz, as it clearly already had on the blues. Most of the surviving heroes of swing and bebop were in decline, and the prevailing wisdom was that no successors were in the wings. But here was an immensely gifted musician still in his teens who played straight-ahead jazz, not fusion or funk. That he was a black second-generation jazzman (the son of the obscure pianist Ellis Marsalis) from New Orleans, the city generally acknowledged to be the birthplace of jazz, was taken as an especially good omen. This was in 1980, and Marsalis has since achieved a celebrity rare for a contemporary jazz musician, partly as a result of his parallel career in classical music, partly as a result of the novelty value of his youth. But at twenty-six, he's no longer a prodigy. He's already three years older than Miles Davis was when he recorded the first of his classic nonet sessions, two years older than Clifford Brown was when he recorded "Jordu" and "Joy Spring." It's time to ask if Marsalis has fulfilled his potential, what his influence has been on musicians his own age and younger, and whether he's expanded the audience for jazz, as many hoped he would.

The answer to the first part of the question is a qualified yes. With his chill tone and jabbing attack, Marsalis still echoes Miles Davis, just as he did eight years ago with Blakey. Moreover, now that Marsalis leads his own band, he takes his cues from the quintet that Davis led from 1964 to 1967, which included the pianist Herbie Hancock, the bassist Ron Carter, the drummer Tony Williams, and the tenor and soprano saxophonist Wayne Shorter (still the primary role model for Marsalis's brother, Branford, Wynton's elder by one year and a former member of his group). So Marsalis is still feeling the Anxiety of Influence, but he's making progress toward ridding himself of it. There's a sly wit to his half-valve work that owes nothing to Davis, although it does recall the veteran trumpeter Clark Terry, one of Davis's early influences. Another sign of Marsalis's enhanced individuality is the way he's zeroed in on the most provocative aspect of the Davis Quintet's music from the middle sixties: its circling, now-you-hear-it-now-you-don't approach to the beat.

At this point, Marsalis is one of very few musicians able to line up enough work to keep a band together fulltime, and his rapport with his rhythm section on the recent *Marsalis Standard Time—Volume 1* (Columbia FC-40461) attests to the virtues of stability. There are moments when Marsalis, the pianist Marcus Roberts, the bassist Robert Leslie Hurst, and the drummer Jeff Watts sound as though they're playing in four different time signatures. Actually, they're stretching a basic quadruple meter four different ways, accenting different beats in every measure, and trusting that the listener will feel the downbeat in his bones. That the listener generally does is a tribute to Watts's zesty drumming, which, like that of Tony Williams with Davis, imparts a sensation of four/four swing while scrupulously avoiding anything so simple as a sounded beat. The effect is mesmerizing, and it would be beyond the ken of a group hastily assembled for a recording date.

Discounting a best-forgotten 1984 encounter with strings, *Marsalis Standard Time—Volume 1* is the first of Marsalis's albums mostly given over to vintage pop songs of the sort that provided excellent springboards for improvisers from Louis Armstrong to John Coltrane, but that too many subsequent musicians have rejected in favor of their own compositions. The album is a reminder of the outstretched hand that such songs have long offered to audiences trying to find a point of entry into jazz. Although the greased tempos frequently reduce the melodies to unrecognizable blurs, the hint of familiarity left in the standards "Caravan," "April in Paris," "A Foggy Day," "The Song Is You," and "Autumn Leaves" brings the quartet's rhythmic cunning into a sharp focus lacking in his other albums.

Marsalis Standard Time—Volume One has its minor flaws. Two Marsalis originals—a fleet blues and an almost motionless ballad—seem out of place, because, surprisingly, they're not subjected to as many rhythmic variations as the standards. "Memories of You," a solo feature for Roberts, begins promisingly, with wrinkled blues shadings and Monk-like rhythmic hesitations, but falters as a result of Roberts's attempt to cram the entire history of jazz piano, from stride to Tyner, into one three-minute performance. There are two versions of "Cherokee," both taken at a tempo considerably less punishing than the one that Charlie Parker set when he reharmonized the song as "Ko Ko," in 1945, but punishing enough to tongue-tie Marsalis. He redeems himself, though, with a lovely interpretation of Benny Goodman's old sign-off theme, "Goodbye," and a straightforward reading of "New Orleans" that comes across as a modernist's heartfelt tribute to Louis Armstrong.

The question of Marsalis's influence is tricky. Even without his example, musicians in their twenties might be looking back two decades for inspiration, to the period before John Coltrane's death in 1967, and the gradual defection of Miles Davis and his sidemen to high-tech funk—the last time when there was anything approaching general agreement on what constituted the state of the art. But inasmuch as Marsalis's emergence identified this anachronistic impulse as a movement, such subsequent arrivals as Branford Marsalis, the alto saxophonist Donald Harrison, the drummer Marvin "Smitty" Smith, and the trumpeters Terence Blanchard and Wallace Roney are Wynton's Children.

Despite his large stylistic debt to Wayne Shorter, Branford Marsalis once seemed the most promising of these young musicians, but he's failed to deliver. His mastery of his instrument is impressive, and so is his grasp of jazz history. Called on to evoke Ben Webster's burly savoir faire on "Take the 'A' Train," on Mercer Ellington's *Digital Duke* (GRP GR-1038), Marsalis performs the task with admirable conviction. But *Renaissance* (Columbia FC-40711), his own most recent album, suggests that his ability to mimic different styles isn't necessarily an asset. Each of his phrases seems to be enclosed in quotation marks: not only "Wayne Shorter" but "John Coltrane," "Sonny Rollins," and "Joe Henderson," too. What's missing is "Branford Marsalis."

Donald Harrison and Terence Blanchard are from New Orleans, like the Marsalis brothers, whom they succeeded in the Jazz Messengers. The recent *Crystal Stair* (Columbia FC-40830) is typical of the four albums they've made as leaders of their own group since 1983. The level of musicianship is high, and there's as much rhythmic detail and acceleration of tempo as on *Marsalis Standard Time—Volume 1*, but Harrison and Blanchard lack Wynton Marsalis's command of dynamics, and the result is that each track is indistinguishable from the one before it. The Marsalis brothers are the models for Harrison and Blanchard, which means that *Crystal Stair* sounds like the 1965 Miles Davis Quintet, once removed. On *Eric Dolphy and Booker Little Remembered Live at Sweet Basil* (ProJazz CDJ-640, available only as a compact disc with a companion cassette of the same material), Harrison and Blanchard are miscast in the roles of two late musicians who were among the most individualistic in jazz. The other participants in this 1986 New York club session are the pianist Mal Waldron, the bassist Richard Davis, and the drummer Ed Blackwell, all reprising the roles they played on the original Dolphy and Little recordings, recorded at the Five Spot in New York in 1961. Although Blanchard is beyond his depth trying to capture Little's melancholy, Harrison surprises the listener with searing improvisations that are satisfying in their own right, even if a little too conventional in pitch to evoke Dolphy.

Marvin "Smitty" Smith, once the drummer in the Harrison–Blanchard group, makes his debut as a leader with *Keeper of the Drums* (Concord Jazz CJ-325), an album that owes much of its vibrancy to Smith's smarts in recognizing the individual abilities of the members of his ensemble—Roney, the trombonist Robin Eubanks, the pianist Mulgrew Miller, the bassist Lonnie Plaxico, and the saxophonists Steve Coleman and Ralph Moore—and allocating solo space in a way that shows each off to best advantage. Coleman's angular phrasing is well suited to the Dolphyesque "Miss Ann," for example, and the sanctified call-and-response patterns of "The Creeper" are custom-made for the blues-based styles of Roney and Miller. Roney's *Verses* (Muse MR-5335), which benefits from Tony Williams's tidal-wave drumming, is more of a blowing date, and this proves to be an advantage on "Slaves" and the title track, two themeless blues with crescendoing solos by Roney, Miller, and the tenor saxophonist Gary Thomas. But on the remaining tracks, including Miles Davis's and Bill Evans' "Blue in Green," the soloists adhere too timidly to the guidelines set by Davis, Evans, and Coltrane on *Kind of Blue*.

On all these recent albums, rhythm is secondary to pulsation, and harmony is frequently suspended in the interest of mood. In that sense, these albums recall those Miles Davis made for Columbia in the middle sixties, and those his sidemen made for Blue Note during the same period, with a stable of likeminded musicians that included the vibraphonist Bobby Hutcherson, the trumpeter Freddie Hubbard, and the tenor saxophonist Joe Henderson. What gave those Blue Note albums urgency was their insistence on moderation at a time when a revolution was going on elsewhere in jazz, in the more iconoclastic music of Coltrane, Ornette Coleman, Cecil Taylor, Albert Ayler, Sun Ra, and Archie Shepp.

These newer records convey urgency, too, but it's the urgency of fighting the clock, of insisting that adventure can still be found in a twenty-year-old style of jazz that represented moderation even when new. It's not surprising these musicians in their twenties are finding a ready audience among lifelong jazz fans at least a decade older. Nostalgia for the recent past is a widespread vice, catered to by the very technology that was supposed to hurl us into a future from which there would be no looking back—the new ghosts in the machine include oldies radio, television reruns, vintage films on videocassette, and Beatles and Motown reissues on compact disc (a format still so expensive that consumers feel safer sticking with the tried and true). But nostalgia is a vice to which jazz fans are especially susceptible. In the late 1960s, Bob Dylan and the Beatles won for rock 'n' roll more intellectual cachet than jazz had ever enjoyed, and both rock and rhythm 'n' blues supplanted jazz as music for hedonistic release. For many listeners, New Age music now serves as the backdrop for meditation that Coltrane and Pharoah Sanders once provided (which is why New Age has been replacing jazz as late-night fare on public radio). It's been a long time, in other words, since a passion for jazz was regarded as hip rather than quaint. So a yearning for a time when Miles Davis was a trendsetter both musically and sartorially is understandable, even among those too young to remember such a time firsthand.

The underside of this nostalgia is a widely felt anger that discounts the onslaught of rock in order to blame jazz's fall on its own excesses after 1965. Wynton Marsalis gives voice to this animosity in statements like the following, from a recent interview with the critic Leonard Feather: "When you come to New York, there's a whole school of musicians who are called the avant-garde, and you don't really [need] any craft requirements to join their ranks. All you have to do is be black and have an African name. . . ." Were it not for the fact that he himself has no past deeds to recant, Marsalis could be the spokesman for the jazz auxiliary of Second Thought, the coterie of *mea culpa*-ing former radicals turned neoconservatives, who blame all of America's current problems on the presumed moral laxity of 1960s. Although Marsalis has expressed admiration for Ornette Coleman, the avant-garde's erosion of standards is a recurring theme in his interviews, and because he neglects to name names, he tars all of Coleman's progeny with the same brush. "You have young musicians who don't know how to play the blues, who don't care about being in tune, who can't get through any of the music that Monk wrote but try to pretend that they're what's going on because they're playing right now," he complained to Feather. And to Stanley Crouch, once an avant-garde drummer and firebrand poet, but now the jazz critic most in sympathy with Marsalis's reactionary aesthetic, Marsalis said, "It's much, much easier to whip up this hasty, fast-food version of innovation than to humble yourself to the musical logics that were thoroughly investigated by [the] masters." To Marsalis and Crouch, free jazz was as costly a mistake as black English; the code of professionalism they espouse is cousin to the rhetoric of the neo-con academics who blame open admissions for the closing of the American mind.

A good many jazz listeners agree with Marsalis, and much of what he says has the ring of truth. "Just to think of the arrogance behind a statement like, 'I play world music . . .'" he told Feather. "You're admitting that you're giving non-specific, second-hand treatment to different types of music. . . ." The avant-garde's naïve fascination with ethnic music is worthy of Marsalis's ridicule. But it's good to remember that in addition to African thumb piano and doussn' gouni and didjeridoo, the avant-garde restored clarinet and tuba to the jazz ensemble, to say nothing of importing such suspect "concert" instruments as violin and cello, thus relieving the inherent monotony of trumpet, saxophone, piano, bass, and drums (the lineup still generally favored by Marsalis and his followers). It was also such nominal avant-gardists of the 1970s as Roscoe Mitchell, Joseph Jarman, Muhal Richard Abrams, Anthony Braxton, Anthony Davis, and Henry Threadgill who rekindled interest in composition (and thus weeded out the amateurs) by avoiding theme-solos-theme formats, and who put jazz back in touch with its pre–Charlie Parker heritage by reinvestigating ragtime, marches, Duke Ellington, Fletcher Henderson, and Jelly Roll Morton.

There's no way to turn back the clock on all that's happened in jazz over the last two decades, nor should we want to. To his credit, Marsalis *has* brought new audiences to jazz, although the critic Steve Futterman isn't the only one wondering if all that Marsalis has accomplished is to persuade "his upscale audiences that jazz could be as boring as they'd always secretly feared." Those of us who are more familiar with the rich diversity of contemporary jazz know that boredom isn't a danger so long as the music keeps evolving. Still, it gives me pause to consider that all of the most intrepid jazz experimentalists are now in their forties or older, while the leading musicians under thirty see themselves as craftsmen making small refinements on a time-tested art. Progress is frequently a myth in jazz, as in most other aspects of contemporary life. But it's a myth so central to the romance of jazz that the cost of relinquishing it might be giving up on jazz altogether.

(April, 1988)

INDEX

258 INDEX

Padgett, Ron, 237
Palmer, Don, 48
Pareles, Jon, 239
Parker, Charlie, ix, 28, 30, 47, 48, 58, 88,
    90, 94, 95n, 96, 102, 139, 158, 161–62,
    164, 166, 194, 197, 199, 201, 203, 211,
    214–22, 223, 238, 240, 243, 246
Parker, Dorothy, 172
Parker, Errol, ix, 16, 53–58
Parker, Evan, 105
Parker, William, 44
Patrick, Pat, 26
Patton, Big John, 73
Paudras, Francis, 212
Peer, Beverly, 155
Penderecki, Krzysztof, 229
Penn, Irving, 92
Pennebaker, D. A., 66n
Peplowski, Ken, 175
Pepper, Art, 98–99, 203
Pepper, Laurie, 99
Peress, Maurice, 3, 4, 12–13, 14, 15, 188–
    90, 200, 204–5
Persip, Charli, 31, 201
Peterson, Oscar, 41, 87, 89, 112
Pettiford, Oscar, 35
Plaxico, Lonnie, 244
Pleasants, Henry, 213
Pleasure, King, 175, 218
Plummer, Paul, 163
Poirier, Richard, 24
Ponomarev, Valery, 228
Ponomareva, Valentina, 233
Pope, Odean, 107–10
Porter, Cole, 116, 138, 150, 154, 155, 168,
    172, 191, 194, 195
Porter, Lewis, 85
Potts, Steve, 104–6
Pound, Ezra, 45, 239
Powell, Bud, 41, 58, 115, 119, 198, 211,
    214
Pozo, Chano, 56
Prado, Perez, 82
Preiss, Jeff, 226
Preservation Hall Jazz Band, 101
Presley, Elvis, 15, 146, 151, 187
Pressley, Julian, 110

Previn, Andre, 143
Previte, Bobby, 73
Price, Sammy, 83–84
Prima, Louis, 177
Prince, 89, 93
Procope, Russell, 9
Professor Longhair, 171
Prokofiev, Sergei, 102
Prymus, Ken, 15
Pryor, Richard, 220

Quebec, Ike, 39
Quinichette, Paul, 88
Quine, Robert, 72

Ra, Sun. See Sun Ra
Rachmaninoff, Sergei, 188, 230
Raeburn, Boyd, 106
Rainger, Ralph, 193
Ramey, Gene, 87
Ravel, Maurice, 119
Ray, Nicholas, 212–13n
Raye, Don, 203
Razaf, Andy, 154
Reagan, Ronald, 180, 183–84
Redgrave, Vanessa, 182
Redman, Don, 25, 67
Reed, Lucy, 33
Reed, Sam, 107
Reese, Della, 8
Reich, Steve, 102, 229
Reid, Vernon, 73
Reilly, Jack, 162
Reinhardt, Django, 54, 198, 228
Richardson, Chan, 211, 215, 217–18, 222
Richardson, Jerome, 206
Riddle, Nelson, 143
Ritter, Clare, 117
Roach, Max, 19, 37, 38, 88, 98, 108, 109,
    199, 203, 214, 221
Roberts, Robbie, 32
Robin, Leo, 155, 168, 193
Rochester, Cornell, 110
Rodgers, Richard, 138, 150, 168, 191,
    193–94, 195
Rodney, Red, 218
Rodrigo, Joaquin, 29

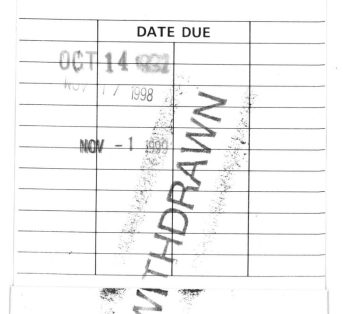

	DATE DUE		
	OCT 14		
	NOV 1 7 1998		
	NOV - 1		